AS Mathematics
UNIT 2

D1323407

Unit 2: Pure Core 2

Steve Walters

Philip Allan Updates, an imprint of Hodder Education, part of Hachette Livre UK, Market Place, Deddington, Oxfordshire OX15 0SE

Orders

Bookpoint Ltd, 130 Milton Park, Abingdon, Oxfordshire OX14 4SB
tel: 01235 827720
fax: 01235 400454
e-mail: uk.orders@bookpoint.co.uk
Lines are open 9.00 a.m.–5.00 p.m., Monday to Saturday, with a 24-hour message answering service. You can also order through the Philip Allan Updates website: www.philipallan.co.uk

ISBN 978-0-340-95749-3

First printed 2008
Impression number 5 4 3 2 1
Year 2013 2012 2011 2010 2009 2008

This guide has been written specifically to support students preparing for the AQA AS Mathematics Unit 2 examination. The content has been neither approved nor endorsed by AQA and remains the sole responsibility of the author.

Typeset by Pantek Arts Ltd, Maidstone
Printed by MPG Books, Bodmin

Hachette Livre UK's policy is to use papers that are natural, renewable and recyclable products and made from wood grown in sustainable forests. The logging and manufacturing processes are expected to conform to the environmental regulations of the country of origin.

Contents

Introduction

■ ■ ■

Content Guidance

■ ■ ■

Questions and Answers

Introduction

About this guide

This unit guide is the second in a series of four which cover the core pure mathematics content of the AQA AS and A2 mathematics specification. It is designed to help you prepare for the **AS Unit 2: Pure Core 2** examination.

The **Introduction** provides advice on how to use the guide, together with suggestions for effective revision.

The **Content Guidance** section gives a summary of the techniques covered in the specification, together with worked examples and exercises that provide opportunities to practise essential skills.

The **Question and Answer** section contains the sort of questions you can expect in the end-of-unit examination. These are meant to give you a flavour of the style and possible content of such questions — not to cover every possibility. Each question is followed by the responses of two candidates. Their answers and the examiner comments that follow should help to guide you towards a good mark while avoiding any pitfalls. This section ends with a **Quick Test** that you should try to complete within 10 minutes.

Using this guide

This guide lends itself to a number of uses throughout the course. It is not just a final revision aid.

The topics in the Content Guidance section follow those in the specification. This means that you can use the Content Guidance section:
- to cross-reference your own notes with the relevant sections here
- as a reference point for homework or assessment tests
- to identify basic strengths and weaknesses, by completing the exercises
- to prepare revision cards or other revision material to summarise the key concepts

The Question and Answer section can be used to:
- identify the terms and language used by examiners
- highlight common errors and learn how to avoid them
- test yourself on several topic areas
- understand the types of mark used by examiners
- identify ways to earn marks

A word of warning — this guide does *not*:
- cover every detail

- use every possible type of question for practice
- go into detail of every possible method or technique that could be applied
- cover concepts presented in unfamiliar situations

Preparing for the examination

Know about the paper

The Pure Core 2 examination contains eight or nine questions. There are short questions worth 4 or 5 marks and longer questions worth up to 18 marks. Many questions fall into the middle mark range. All questions must be attempted.

The exam lasts 90 minutes and there are 75 marks available. This means you can spend, at most, $1\frac{1}{5}$ minutes per mark. This is crucial to giving you an idea of pace when tackling the paper, e.g. 5 marks = 6 minutes.

A standard calculator is allowed — scientific or graphical. It must *not* be an algebraic manipulator.

The standard formulae book is available but contains a limited number of data applicable to this module. Take no chances — make sure that you learn all the key formulae. The specification assumes a good knowledge of techniques from Pure Core 1.

This paper makes up one-third of the total AS marks or one-sixth of the marks for A-level.

Command words

Examiners use command words to indicate the skills that are being tested. If you are to be successful, you must know what these words mean. Common command words used in Pure Core examination papers are given below, together with brief explanations of what is required.

Approximate
A numerical method must be used, as an exact answer is not possible or desirable.

By considering
This is asking you to use a particular method of solution.

Calculate
Calculator work is required to achieve a numerical answer, possibly with a stated range of accuracy.

Describe
This requires some explanation in the form of text, possibly supported by working out.

Determine

This is asking you to provide full working out in order to establish a particular result or feature. Such questions are often worth a number of significant marks. It is more discriminating than other command words.

Express

This is used in relation to algebraic expressions, including surds. A particular form is required for the answer. The examiner is hinting at the structure of the answer that is expected.

Find

This instruction is used in a variety of questions, which are usually worth 2 or 3 marks. It means calculate or work out.

Hence, or otherwise

The word 'hence' indicates that the answer required follows on from the preceding part of the question. 'Or otherwise' allows you to choose another route through the problem. However, you should be aware that an alternative route may take longer to reach the answer.

Prove

This instruction is used in algebraic questions involving the rules of algebra along with trigonometric identities or laws of logarithms. Your first line of working should start with one expression and your last line of working should match the required expression.

Show that

The question gives a result that you have to show is true. If you can do this, then you know that you will have gained the marks.

Simplify

This instruction is used in relation to algebraic expressions, including surds. The idea is to rewrite the expression in a different form — for example, by expanding brackets, collecting terms or using identities.

Sketch

This command word is used in relation to sketching curves. You should indicate the general shape of the curve and identify key points, such as intersections with axes and stationary points (if found).

Solve

This term is used in relation to equations and inequalities. It means that you have to find all solutions that satisfy a particular equation or give a range of values that satisfy an inequality.

State

A quick answer, worth 1 or 2 marks, is required. Any working is likely to be done in your head.

Verify

This requires you to show that a stated answer or property applies. The given information is used to show that it is true.

Write down

A quick answer, worth 1 or 2 marks, is required. The answer should be obvious.

When simplifying, proving or showing, remember use the 'four Fs'. These are:

- **F**actorise. Can the expression be factorised? Common factor, difference between two squares or trinomial approaches?
- **F**raction. Can the rules of fractions be used to simplify — add, subtract, multiply, divide, cancel?
- **F**OIL. A mnemonic to remember how to expand brackets (First, Outside, Inside, Last). Can the brackets be expanded?
- **F**ind an identity. Usually a trigonometrical one, such as $\tan \theta = \frac{\sin \theta}{\cos \theta}$ or $\cos^2 \theta + \sin^2 \theta = 1$.

Revision

Organising your revision

Your revision folder should contain a mixture of notes and practice questions. It is a good idea to summarise key points on revision cards. These should give the main results for a topic with examples of their use. The Content Guidance section of this guide should help you to do this. You could clip the cards together to make a booklet that you can carry with you for revision.

It is important to plan your time and start your revision early. If you are taking the Pure Core 2 examination in January, start revising in the October half term. If you are sitting the examination in the summer, start revising before Easter. Aim to revise in daily sessions lasting 30–45 minutes. Use the topic list provided to allocate revision time effectively.

Remember that the content of Pure Core 2 develops and extends several ideas from Pure Core 1. It may well be worth revisiting such topics as differentiation and integration in the Pure Core 1 module.

Leave practising answering questions from past papers until you are confident. You can test yourself by theme, picking questions out from several papers, or try to do a whole paper in the time allowed.

Analysing past papers

The following tables give information about the topics tested and the number of marks allocated to each topic over the past seven examinations. In each table, the final column gives the percentage allocation out of the total marks available (525) over the seven papers.

You can use these tables to help plan your revision. However, you should be aware that the list is not exhaustive and it can be a matter of opinion how to classify some marks. As you consider each table, you should bear in mind these questions:

- What proportion of your revision time should you allocate to each main section?
- Which topics occur on (almost) every paper?
- Which topics are worth the most marks? Why?
- Which topics are least likely to be tested?
- Are there any patterns in the frequency of certain topics being tested?
- How do you rate yourself on each topic? Use a 'traffic lights' scheme — red, amber or green.

Algebra and functions

Topic	Jan 08	Jun 07	Jan 07	Jun 06	Jan 06	Jun 05	Jan 05	Total	%
Laws of indices		3	5	1		3	3	15	2.9
Description of transformation		2	2	4	6	2	2	18	3.4
Transformations — forming expressions	4	1		2			1	8	1.5
								41	8

- This topic accounts for 8% of the marks.
- Transformations are tested more often than laws of indices.

Sequences and series

Topic	Jan 08	Jun 07	Jan 07	Jun 06	Jan 06	Jun 05	Jan 05	Total	%
Sequences — using the nth term formula		2				2		4	0.8
Sequences — recurrence relations					6			6	1.1
Recurrence relations — limits					3			3	0.6
Arithmetic series — using formulae	5	7		2		4	4	22	4.2

Topic	Jan 08	Jun 07	Jan 07	Jun 06	Jan 06	Jun 05	Jan 05	Total	%
Arithmetic series — forming equations				5			4	9	1.7
Geometric series — using formulae		5				4		9	1.7
Geometric series — forming equations			5					5	1
Sum to infinity			2		6	3		11	2.1
Binomial expansions	10	2	4	3		6	7	32	6.1
Binomial coefficients			3	5				8	1.5
								109	21

- This topic accounts for the second largest proportion of marks (21%).
- Series questions, which have the highest proportion of marks, are varied.
- Arithmetic and geometric series are rarely tested together.
- Binomial expansion techniques can also focus on obtaining coefficients.

Trigonometry

Topic	Jan 08	Jun 07	Jan 07	Jun 06	Jan 06	Jun 05	Jan 05	Total	%
Sine rule	3			3				6	1.1
Cosine rule		3	3		3	3	3	15	2.9
Area of triangle — sine formula	3	3	2	3	3	2	3	19	3.6
Arc length or perimeter of sector	3	2	3	3	2	3	2	18	3.4
Sector area	3	2	2	2	3	2	2	16	3
Graphs of sine, cosine, tangent		3	3				4	10	1.9
Solving simple trigonometrical equations	4	2	7	4	3	2	5	27	5.1
Solving equations using identities		5		5		5		15	2.9
Proving identities/ using identities	4		3		4	3		14	2.7
								140	27

- This topic accounts for largest proportion of marks (27%).
- The cosine rule is tested more often than the sine rule.
- The use of arc and sector area formulae is standard.
- The solution of trigonometrical equations has the highest proportion of marks.

Exponentials and logarithms

Topic	Jan 08	Jun 07	Jan 07	Jun 06	Jan 06	Jun 05	Jan 05	Total	%
Exponential graphs/values	2	3		3			1	9	1.7
Logarithms — solving equations	1	3	2	3		2	3	14	2.7
Logarithms — manipulating expressions	3	5	9	3	5	6	4	35	6.7
Solving exponential equations	6	3		4	3		4	20	3.8
								78	15

- This topic accounts for 15% of the marks.
- Manipulation using the laws of logarithms, particularly in relation to algebraic expressions, has the highest proportion of marks.
- Various types of logarithmic equations may be tested.
- Only simple exponential equations are tested, although the 'quadratic' type may occur with hints for their solution.

Differentiation

Topic	Jan 08	Jun 07	Jan 07	Jun 06	Jan 06	Jun 05	Jan 05	Total	%
Differentiating powers of x	3	3	3	3	2	2	3	19	3.6
Finding gradients — substituting	2	2		2	1		1	8	1.5
Finding equation of a tangent					7	4		11	2.1
Finding equation of a normal	3	5	4	3		3	4	22	4.2
Finding stationary points	3		4	2	5		3	17	3.2
Application, e.g. lines and intersections	3							3	
Increasing functions						2		2	0.4
								82	16

- Differentiation accounts for 16% of the marks.
- This topic mirrors the one in Pure Core 1, but now applies the techniques to negative and fractional powers of *x*.
- The use of indices to rewrite expressions has the highest proportion of marks.
- Finding the equation of the normal is tested far more often than finding the equation of the tangent.
- Applications may use knowledge of coordinate geometry to find intersections and areas.

Integration

Topic	Jan 08	Jun 07	Jan 07	Jun 06	Jan 06	Jun 05	Jan 05	Total	%
Integrating powers of *x*	3	3	3	3	3	2	3	20	3.8
Definite integrals		2					3	5	1
Finding the equation of a curve				3				3	0.6
Integrals and areas	3		2		5	6		16	3
Using the trapezium rule	4	4	4	4	5	4	6	31	5.9
								75	14

- This topic accounts for 14% of the marks.
- The use of the trapezium rule is always tested. Questions that relate to the accuracy of the estimate account for more marks.
- Questions may ask for an integral and then use it, e.g. relate to an area etc.

The examination

- Open the question paper and scan the questions.
- Identify the questions that you feel confident about.
- Start with a question that will boost your confidence — for example, a question on your favourite topic.
- Read the stem of the questions carefully and look for command words.
- Take note of the mark allocation and pace yourself. Remember that 5 marks are equivalent to a maximum of 6 minutes. If there are only a few marks available, your response should be short.
- If you get stuck on a question, do not waste too much time on it. Carry on and come back to it later.

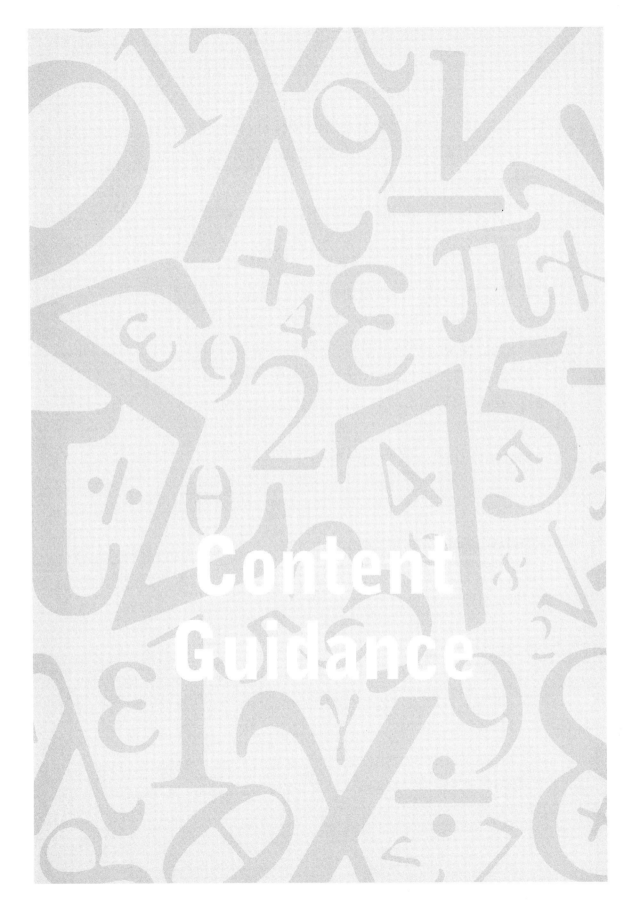

Content
Guidance

This section is a guide to the content of **Unit 2: Pure Core 2**. The material follows the order used in the specification.

Algebra and functions

- Laws of indices
- Transformations on $f(x)$

Sequences and series

- The binomial expansion of $(1 + x)^n$
- Sequences — nth term, iterative sequences, limits
- Arithmetic series
- Geometric series including sum to infinity

Trigonometry

- The sine and cosine rules; area of a triangle — sine form
- Arc length, area of a sector (use of radians)
- Sine, cosine and tangent — graphs, symmetries and periodicity
- Solution of trigonometrical equations — use of standard identities

Exponentials and logarithms

- Graph of $y = a^x$
- Laws of logarithms
- The solution of $a^x = b$

Differentiation and integration

- Differentiation and integration of x^n, where n is negative or fractional
- Definite integrals — application to areas
- Approximation of areas using the trapezium rule

Algebra

Indices

Key learning objectives
- to use index notation with integer, fractional and negative powers
- to use the laws of indices to evaluate numerical expressions
- to use the laws of indices to rewrite algebraic expressions in index form

Index form

Numbers written in the form a^b are said to be in **index form**. The b is called the **index number** and a is referred to as the **base number**. For example with 3^2, the 3 is the base and 2 is the index. The plural of index is **indices**. Scientific calculators have an a^b or x^y button but are of little use for algebraic manipulation. There are a set of rules and special results to learn in order to be able to manipulate or evaluate expressions involving indices, as shown below.

Rules of indices

$a^m \times a^n = a^{m+n}$ (when multiplying, add the indices)

$a^m \div a^n = a^{m-n}$ (when dividing, subtract the indices)

$(a^m)^n = a^{mn}$ (when raising to a power, multiply the indices)

$a^0 = 1$ (anything to the power of 0 is 1)

$a^{-m} = \dfrac{1}{a^m}$ (negative indices mean reciprocals i.e. 'one over...')

$a^{\frac{1}{n}} = \sqrt[n]{a}$ (fractional indices mean nth roots)

$a^{\frac{m}{n}} = \left(a^{\frac{1}{n}}\right)^m = \left(\sqrt[n]{a}\right)^m$ (combining rules for raising powers and roots)

$a^{\frac{-m}{n}} = \dfrac{1}{a^{\frac{m}{n}}} = \dfrac{1}{\left(\sqrt[n]{a}\right)^m}$ (combining rules for raising powers, roots and reciprocals)

Worked example 1

Write each of the following as a single rational number.

(a) 5^{-3} **(b)** $1000^{\frac{2}{3}}$ **(c)** $36^{-\frac{3}{2}}$

Solution

(a) $5^{-3} = \dfrac{1}{5^3} = \dfrac{1}{125}$

(b) $1000^{\frac{2}{3}} = \left(\sqrt[3]{1000}\right)^2 = 10^2 = 100$

(c) $36^{-\frac{3}{2}} = \dfrac{1}{36^{\frac{3}{2}}} = \dfrac{1}{(\sqrt[2]{36})^3} = \dfrac{1}{6^3} = \dfrac{1}{216}$

Worked example 2

Write down the values of a, b and c given that:

(a) $9^a = 3$

(b) $9^b = \dfrac{1}{3}$

(c) $9^c = 27$

Solution

(a) $a = \dfrac{1}{2}$ since 3 is the square root of 9. This is the key to identifying the other values.

(b) $b = -\dfrac{1}{2}$ since $\dfrac{1}{3}$ is the reciprocal of 3.

(c) $c = \dfrac{3}{2}$ since 27 is 3 cubed.

Worked example 3

(a) Simplify $(x^{\frac{3}{4}})^{12}$.

(b) Express $x^2 \sqrt{x}$ in the form x^p.

(c) Express $\dfrac{x^{10} + 1}{x^4}$ in the form $x^a + x^b$ where a and b are integers to be found.

Solution

(a) $\left(x^{\frac{3}{4}}\right)^{12} = x^{\frac{3}{4} \times 12}$. This uses the rule for raising to another power.

(b) $x^2 \sqrt{x} = x^2 x^{\frac{1}{2}} = x^{2+\frac{1}{2}} = x^{\frac{5}{2}}$ so $p = \dfrac{5}{2}$. This uses the rule for fractional indices and then adding indices.

(c) $\dfrac{x^{10} + 1}{x^4} = \dfrac{x^{10}}{x^4} + \dfrac{1}{x^4} = x^{10-4} + x^{-4} = x^6 + x^{-4}$ so $a = 6$ and $b = -4$.

Note that you have to write the expression as two fractions before you can use any rules of indices. Then use the subtraction of indices and reciprocals.

Note that algebraic manipulation of indices is often required in differentiation and integration questions.

Exercise 1

(1) Evaluate $16^{-\frac{1}{2}}$.

(2) Express $\left(\dfrac{49}{16}\right)^{-\frac{3}{2}}$ as a single rational number.

(3) State the value of p given that $4^p = \dfrac{1}{64}$.

(4) Find x given that $5^{-x} = \sqrt{5}$.

(5) Find the value of q such that $\dfrac{5^q}{\sqrt{5}} = \dfrac{1}{25}$.

Exercise 2

(1) Simplify $x^{\frac{1}{3}} \times x^{\frac{2}{3}}$.

(2) Express $x^{\frac{5}{2}} \times x$ in the form x^p.

(3) Write $x\sqrt{x}$ as a power of x.

(4) Express $x\left(\sqrt{x} - 1\right)$ in the form $x^a - x^b$ where a and b are rational numbers.

(5) Express $\dfrac{4x^{12} - 2}{2x^3}$ in the form $ax^b - x^c$ where a, b and c are rational numbers.

Answers to exercises: indices

Exercise 1

(1) $\dfrac{1}{4}$

(2) $\dfrac{64}{343}$

(3) $p = -3$

(4) $x = \dfrac{1}{2}$

(5) $-\dfrac{3}{2}$

Exercise 2

(1) x or x^1

(2) $x^{\frac{7}{2}}$

(3) $x^{\frac{3}{2}}$

(4) $x^{\frac{3}{2}} - x^1$ or $x^{\frac{3}{2}} - x$

(5) $2x^9 - x^{-3}$

Functions

Key learning objectives

- to identify families of curves and the link between them
- to describe a single transformation using either translations, reflections or stretches
- to find an equation for a curve after a transformation has been applied
- to use function notation to identify and learn the six standard transformations

By considering the structure of a function you can identify the family of curves to which it belongs.

For example, $y = x^2 + 4$, $y = (x - 1)^2$, $y = 5 - 3x^2$ all belong to the family of quadratic curves. The simplest such curve is $y = x^2$.

Similarly, $y = \sin 2x$, $y = 1 + 3\sin x$, $y = \sin(x - 30°)$ all belong to the family of sine curves. The simplest such curve is $y = \sin x$.

When in the right format it is possible to link any curve to the simplest form of that family by using a series of transformations. A **transformation** can be a reflection, translation or stretch that is used to alter the position and size of the curve. It is important to remember that in order to define:

- a reflection, you must state the line of reflection
- a translation, you must state a column vector $\begin{bmatrix} a \\ b \end{bmatrix}$ where a is the horizontal movement and b is the vertical movement (right = positive, left = negative; up = positive, down = negative)
- a stretch, you must state the scale factor and direction — either horizontal or vertical

The curves to be transformed will be polynomials, exponentials, trigonometric or rational functions.

To illustrate this idea, consider the following functions, which all belong to the family of cubic curves:

$y = x^3 + 2$ The curve $y = x^3$ has been translated by $\begin{bmatrix} 0 \\ 2 \end{bmatrix}$

$y = (x + 2)^3$ The curve $y = x^3$ has been translated by $\begin{bmatrix} -2 \\ 0 \end{bmatrix}$

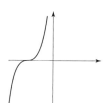

$y = 2x^3$ The curve $y = x^3$ has been stretched vertically by a scale factor of 2

$y = (2x)^3$ The curve $y = x^3$ has been stretched horizontally by a scale factor of $\frac{1}{2}$

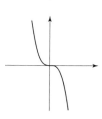

$y = -x^3$ The curve $y = x^3$ has been reflected in the x-axis ($y = 0$)

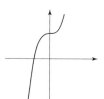

$y = (-x)^3$ The curve $y = x^3$ has been reflected in the y-axis ($x = 0$)

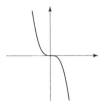

The type of transformation depends on which type of operation has been used in the formula for the function. For example, is a number being added? Is a number being multiplied? Is the operation in the bracket or outside the bracket? You can use this idea to generalise for any type of function.

In order to generalise, you use function notation. The notation $f(x)$ denotes the function f (essentially the 'name') with variable x. To link to specific functions, write $f(x) = x^2$ or $g(x) = \sin x$.

By using just $f(x)$ you do not specify any particular function. The six generalisations are listed in the following table.

Transformation — algebraic	Transformation — descriptive	Comment
$y = f(x) + a$	Translate the curve $y = f(x)$ by $\begin{bmatrix} 0 \\ a \end{bmatrix}$	In the formula we have added a *outside* the function. This results in a **vertical** translation.
$y = f(x + a)$	Translate the curve $y = f(x)$ by $\begin{bmatrix} -a \\ 0 \end{bmatrix}$	In the formula we have added a *inside* the function. This results in a **horizontal** translation. Note that the translation is the *inverse* of '+ a'. Replace x by $x + a$.
$y = a f(x)$	Stretch the curve $y = f(x)$ vertically by scale factor a	In the formula we have multiplied by a *outside* the function. This results in a **vertical** stretch.
$y = f(ax)$	Stretch the curve $y = f(x)$ horizontally by scale factor $\frac{1}{a}$	In the formula we have multiplied by a *inside* the function. This results in a **horizontal** stretch. Note that the scale factor is the *inverse* of '× a'. Replace x by ax.
$y = -f(x)$	Reflect the curve in the x-axis ($y = 0$)	The negative sign is outside the function and the curve has changed **vertically**.
$y = f(-x)$	Reflect the curve in the y-axis ($x = 0$)	The negative sign is inside the function and the curve has changed **horizontally**. Replace x by $-x$.

It is possible to combine a series of transformations but this is excluded from the Pure Core 2 specification. At most one transformation will be required. There are three main types of examination questions:

- Type 1 — those that require a description of a transformation that transforms $y = f(x)$ into one of the above types.
- Type 2 — those that require you to state the new equation for the curve after a given transformation has been applied.
- Type 3 — those that require you to sketch a curve after a given transformation. The transformation will be defined algebraically. In this case the given curve may not even be one that you recognise — it may be abstract.

Worked example

(a) Describe the single geometrical transformation by which the curve with equation $y = \sqrt{x + 2}$ can be obtained from $y = \sqrt{x}$.

(b) Describe the single geometrical transformation by which the curve with equation $y = 5(3^x)$ can be obtained from $y = 3^x$.

Solution

(a) Note that the '+ 2' occurs *inside* the function. This tells us that the transformation is a horizontal translation.

The correct description is translate the curve $y = \sqrt{x}$ by $\begin{bmatrix} -2 \\ 0 \end{bmatrix}$.

(b) Note that the '× 5' is *outside* the function. This means that the transformation is a vertical stretch. The correct description is stretch $y = 3^x$ vertically by scale factor 5.

Exercise 1

(1) Describe the single geometrical transformation by which the curve with equation $y = \sin 2x$ can be obtained from $y = \sin x$.

(2) Describe the single geometrical transformation by which the curve with equation $y = (x + 1)^5 - 3$ can be obtained from $y = x^5$.

(3) Describe the single geometrical transformation by which the curve with equation $y = 4^{-x}$ can be obtained from $y = 4^x$.

Worked example

(a) The curve $y = 1 + \sin 2x$ is reflected in the x-axis to give the curve with equation $y = f(x)$. Write down an expression for $f(x)$.

(b) The curve $y = x^4$ is translated $\begin{bmatrix} -2 \\ 3 \end{bmatrix}$ to give the curve with equation $y = f(x)$. Write down an expression for $f(x)$.

Solution

(a) In order to reflect in the x-axis you must put a negative sign *outside* the function. Therefore $f(x) = -(1 + \sin 2x)$ or $-1 - \sin 2x$.

(b) This time the translation is both horizontally and vertically. The '–2' affects the *inside* of the function, while the '3' affects the *outside* of the function. Therefore, $f(x) = (x + 2)^4 + 3$.

Exercise 2

(1) The curve $y = \cos 3x$ is translated $\begin{bmatrix} 0 \\ -2 \end{bmatrix}$ to give the curve with equation $y = f(x)$. Write down an expression for $f(x)$.

(2) The curve $y = 10^x$ is stretched horizontally by scale factor $\frac{1}{7}$ to give the curve with equation $y = f(x)$. Write down an expression for $f(x)$.

(3) The curve $y = 5 - x^3$ is reflected in the y-axis to give the curve with equation $y = f(x)$. Write down an expression for $f(x)$.

Worked example

The graph below shows the curve $y = f(x)$.

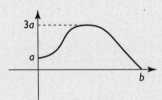

Sketch each of the following graphs on separate axes.

(a) $y = f(x) - a$

(b) $y = 2f(x)$

Solution

(a) The 'a' has been subtracted 'outside' the function therefore the graph is translated by $\begin{bmatrix} 0 \\ -a \end{bmatrix}$ i.e. vertically down by 'a'.

Therefore the graph looks like this:

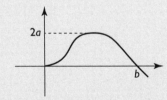

(b) The '2' has been multiplied outside the function therefore the graph is stretched vertically by a scale factor 2. The graph therefore looks like this:

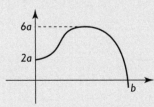

Exercise 3

The graph of $y = f(x)$ is shown below.

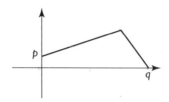

Sketch each of the following graphs on separate axes:

(a) $y = -f(x)$

(b) $y = f(x + q)$

(c) $y = f(x) + p$

Answers to exercises: functions

Exercise 1

(1) Horizontal stretch by scale factor $\frac{1}{2}$

(2) Translate by $\begin{bmatrix} -1 \\ -3 \end{bmatrix}$

(3) Reflect in the y-axis

Exercise 2

(1) $y = \cos 3x - 2$ or $y = -2 + \cos 3x$

(2) $y = 10^{7x}$

(3) $y = 5 + x^3$

Exercise 3

(a)

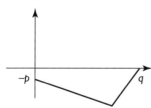

(c)

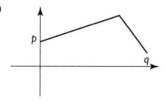

(b)

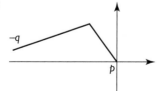

Sequences and series

Binomial expansions

Key learning objectives
- to be aware of Pascal's triangle and use it to expand brackets
- to understand and use $\binom{n}{r}$ notation
- to expand $(x + y)^n$ for integer values of n
- to use the structure of coefficients to efficiently solve a range of problems involving binomial coefficients

The pattern of numbers shown below is part of Pascal's triangle. The numbers in each row are used in several areas of mathematics — in particular with probability and algebraic expansions. In Core 2 you need to use these numbers to expand brackets.

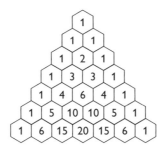

Note that the triangle is symmetrical. The number 1 starts and ends each row. The numbers in subsequent rows are made by adding together the two numbers immediately above. So the next row would be:

1, 1 + 6 = 7, 6 + 15 = 21, 15 + 20 = 35 then repeat in reverse to get 35, 21, 7, 1

This triangle has a connection with powers of 11 — can you see what it is? Why does it then stop? (or does it?)

To generate any more rows you can continue the pattern or use a formula or a special calculator function. To refer to individual coefficients on each row there is a special notation. The **(r + 1)th** number on **row n** is denoted by $_nC_r$ or ^nC_r or $\binom{n}{r}$. This is often referred to as 'choosing r items from n objects'. The AQA specification uses the latter notation. Calculators tend to use one of the first two — either above the number keys or as a special function, usually in a Statistics option. For example, the numbers in row 4 of Pascal's triangle are denoted as follows:

$$_4C_0 = {}^4C_0 = \binom{4}{0} = 1, \; _4C_1 = {}^4C_1 = \binom{4}{1} = 4, \; _4C_2 = {}^4C_2 = \binom{4}{2} = 6, \; _4C_3 = {}^4C_3 = \binom{4}{3} = 4,$$

$$_4C_4 = {}^4C_4 = \binom{4}{4} = 1$$

To use the calculator, enter the value of n, press $_nC_r$, enter the value of r, then press =.

So for $\binom{7}{3}$, $n = 7$ and $r = 3$ therefore $_7C_3 = 35$.

Alternatively the formula for $\binom{n}{r} = \dfrac{n!}{r!(n-r)!}$ where $n!$ denotes **n factorial** and

$n! = n \times (n-1) \times (n-2) \times \ldots \times 3 \times 2 \times 1$ (the product of all integers less than or equal to n, and $0! = 1$).

> ### Worked example
>
> For $\binom{7}{3} = \dfrac{7!}{3!4!} = \dfrac{7 \times 6 \times 5 \times 4 \times 3 \times 2 \times 1}{(3 \times 2 \times 1)(4 \times 3 \times 2 \times 1)} = 35$ (notice that many
>
> numbers cancel).

Exercise 1
Find the values of:

(a) $\binom{6}{4}$

(b) $\binom{10}{5}$

(c) the first four coefficients in row 11 of Pascal's triangle

To see how these coefficients are used with expansions, consider the following:

$(x + y)^1 = \qquad\qquad \mathbf{1}x + \mathbf{1}y$

$(x + y)^2 = \qquad \mathbf{1}x^2 + \mathbf{2}xy + \mathbf{1}y^2$

$(x + y)^3 = \quad \mathbf{1}x^3 + \mathbf{3}x^2y + \mathbf{3}xy^2 + \mathbf{1}y^3$

$(x + y)^4 = \mathbf{1}x^4 + \mathbf{4}x^3y + \mathbf{6}x^2y^2 + \mathbf{4}xy^3 + \mathbf{1}y^4$

Note the following observations:
- The coefficient of each term appears in the corresponding row of Pascal's triangle. Therefore **row n** in Pascal's triangle is used when expanding $(x + y)^n$.
- The degree of each term in the **nth row = n**. For example, all terms in the third row have degree 3 (x^3, x^2y, xy^2 and y^3).
- After expanding $(x + y)^n$ the powers of x decrease from n to 0, while the powers of y increase from 0 to n.
- Therefore, each term on row n has the structure $\binom{n}{r} x^{n-r} y^r$.

Using the above observations you could write a general formula:

$(x + y)^n =$

$$\binom{n}{0}x^n + \binom{n}{1}x^{n-1}y^1 + \binom{n}{2}x^{n-2}y^2 + \ldots + \binom{n}{r}x^{n-r}y^r + \ldots + \binom{n}{n-1}x^1 y^{n-1} + \binom{n}{n}y^n$$

Note that this formula becomes easy to use if either x or y has the value 1, since $1^n = 1$ for all values of n.

Essentially two main types of question are asked:
- Those that require one or more brackets to be expanded fully and simplified. Additionally, specific values may need to be substituted. Use the general formula or Pascal's triangle to answer these questions (see Exercise 2 on p. 26).
- Those that require a specific coefficient to be obtained — possibly involving additional multiplication of part of a bracket. The best way to approach these is to use the observations made above and think about the terms you get when expanding brackets (see Exercise 3 on p. 27).

Below are two examples of each type.

Worked example 1
Given that $(1 + 2x)^4 = 1 + 8x + px^2 + qx^3 + 16x^4$, find the values of p and q.

Solution
Use row 4 of Pascal's triangle so the numbers are 1   4   6   4   1
Powers of 1 will decrease from $(1)^4$ while powers of $2x$ will increase from $(2x)^0$
This gives:
$1(1)^4(2x)^0 + 4(1)^3(2x)^1 + 6(1)^2(2x)^2 + 4(1)^1(2x)^3 + 1(1)^0(2x)^4$
Note that the 0 powers could be left out but they do help to illustrate the structure involved.
This simplifies to $1 + 8x + 24x^2 + 32x^3 + 16x^4$ therefore $p = 24$ and $q = 32$.
It is important to note the use of brackets to remember to raise the '2' and the 'x' to the required power.

Worked example 2
Simplify fully $(3 + x)^5 - (3 - x)^5$.

Solution
Considering $(3 + x)^5$, you need to use row 5 of Pascal's triangle:
1   5   10   10   5   1
Powers of 3 will decrease from $(3)^5$, while powers of x will increase from $(x)^0$.
This gives:
$1(3)^5(x)^0 + 5(3)^4(x)^1 + 10(3)^3(x)^2 + 10(3)^2(x)^3 + 5(3)^1(x)^4 + 1(3)^0(x)^5$
$= 243 + 405x + 270x^2 + 90x^3 + 15x^4 + x^5$

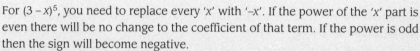

For $(3 - x)^5$, you need to replace every 'x' with '$-x$'. If the power of the 'x' part is even there will be no change to the coefficient of that term. If the power is odd then the sign will become negative.

Therefore, $(3 - x)^5 = 243 - 405x + 270x^2 - 90x^3 + 15x^4 - x^5$

The final expression is $(243 + 405x + 270x^2 + 90x^3 + 15x^4 + x^5) - (243 - 405x + 270x^2 - 90x^3 + 15x^4 - x^5)$

$= 810x + 180x^3 + 2x^5$

Exercise 2

(1) Expand $(1 + x)^5$. Hence express $(1 + \sqrt{2})^5$ in the form $p + q\sqrt{2}$.

(2) Expand $(2 + 5x)^3$. Hence simplify fully $(2 + 5x)^3 - (2 - 5x)^3$.

(3) Given that that $(1 - 3x)^4 = 1 + ax + bx^2 + cx^3 + 81x^4$, find the values of a, b and c. Hence show that:

$(1 - 3x)^4 - (1 + 3x)^4 = -24x - 216x^3$

Worked example 1

Find the coefficient of x^7 in the expansion of $(2 - x)^{13}$.

Solution

Consider the structure of each term. Row 13 of Pascal's triangle is needed.

'$-x$' will be raised to the power of 7 so '2' will be to the power of $13 - 7 = 6$.

The required coefficient from Pascal's triangle will be $\binom{13}{7} = 1716$.

So the required term will be $1716(2)^6(-x)^7 = -109\,824x^7$ and therefore the coefficient is $-109\,824$.

Worked example 2

Find the coefficient of x^5 in the expansion of $(1 + 4x)(1 - 3x)^5$.

Solution

Think first how the x^5 could be made up from the brackets. There are just two possibilities.

Either (constant from the first bracket)(x^5 term from the second bracket) or (x term from the first bracket)(x^4 term from the second bracket).

Considering the first option, from the first bracket you get 1; from the second bracket you get:

$(1)^0(-3x)^5 = -243x^5$. Hence $(1\ldots)(\ldots - 243x^5) = \ldots -243x^5$

Considering the second option, from the first bracket you get $4x$; from the second bracket you get:

$(1)^1(-3x)^4 = 81x^4$. Hence $(\ldots + 4x)(\quad + 81x^4\ldots) = 324x^5$

Hence the required term is $-243x^5 + 324x^5 = 81x^5$ and the coefficient is therefore 81.

Exercise 3

(1) Find the coefficient of x^5 in the expansion of $(1 - 2x)^{10}$. Hence obtain the coefficient of x^5 in the expansion of $(2 + x)(1 - 2x)^{10}$.

(2) Find the coefficient of x^7 in the expansion of $(2 + 3x)^8$. Hence obtain the coefficient of x^7 in the expansion of $(1 - 4x^2)(2 + 3x)^8$.

(3) Find the coefficient of x in the expansion of $(1 + 4x)^7$. Hence find the coefficient of x in the expansion of $(1 + 4x)^7(1 - 2x)^{10}$.

Answers to exercises: binomial expansions

Exercise 1

(a) 15

(b) 252

(c) 1, 11, 55, 165

Exercise 2

(1) $1 + 5x + 10x^2 + 10x^3 + 5x^4 + x^5$ and $41 + 29\sqrt{2}$

(2) $1(2)^3(5x)^0 + 3(2)^2(5x)^1 + 3(2)^1(5x)^2 + 1(2)^0(5x)^3 = 8 + 60x + 150x^2 + 125x^3$ and $(8 + 60x + 150x^2 + 125x^3) - (8 - 60x + 150x^2 - 125x^3) = 120x + 250x^3$

(3) $1(1)^4(-3x)^0 + 4(1)^3(-3x)^1 + 6(1)^2(-3x)^2 + 4(1)^1(-3x)^3 + 1(1)^0(-3x)^4 = 1 - 12x + 54x^2 -108x^3 + 81x^4$ so $a = -12$, $b = 54$ and $c = -108$.

Hence $(1 - 12x + 54x^2 - 108x^3 + 81x^4) - (1 + 12x + 54x^2 + 108x^3 + 81x^4) = -24x -216x^3$

Exercise 3

(1) The term is $\binom{10}{5}(1)^5(-2x)^5 = -8064x^5$ so the coefficient is -8064.

Consider (constant term)(x^5 term) + (x term)(x^4 term).

This gives $(2)(-8064x^5) + (1x)\binom{10}{4}(1)^6(-2x)^4 = -12\,768x^5$.

Coefficient $= -12\,768$

(2) The term is $\binom{8}{7}(2)^1(3x)^7 = 34\,992x^7$ so the coefficient is $34\,992$.

Consider (constant term)(x^7 term) + (x^2 term)(x^5 term).

This gives $(1)(34\,992)x^7 + (-4x^2)\binom{8}{5}(2)^3(3x)^5 = -400\,464x^7$.

Coefficient $= -400\,464$

(3) The term is $\binom{7}{1}(1)^6(4x)^1 = 28x$ so the coefficient is 28.

Consider (constant term)(x term) + (x term)(constant term).

This gives $(1)\binom{10}{1}(1)^9(-2x)^1 + (28x)(1) = 8x$.

Coefficient $= 8$

Types of sequences and series

Key learning objectives

- to use formulae or inductive definitions to generate sequences and where possible to find a limiting value
- to identify the differences between arithmetic and geometric series
- to use formulae for nth terms and sum to n terms to solve a range of problems involving arithmetic and geometric series
- to find the sum to infinity of a geometric series and know the condition for this to be possible

A list of numbers that follow a defined pattern or rule is called a **sequence**. A **series** occurs when terms of a sequence are added together. Some well-known sequences include 2, 4, 6, 8,... (even numbers); 1, 4, 9, 16,...(square numbers); 1, 1, 2, 3, 5, 8,... (Fibonacci sequence). Each individual number in a sequence is called a **term**. To refer to individual terms, use the **suffix notation**, so that u_1, u_2, u_3,... (any letter can be used but 'u' is the most common) would denote the first, second and third terms respectively of a sequence. It follows that $\boldsymbol{u_n}$ would denote the nth term. Clearly since n refers to a position it must always be a positive integer.

Sequences can be defined algebraically in two main ways: **position-to-term rules (nth term formula)** or **term-to-term rules (inductive definition — start value and rule)**. As an example, consider the sequence of square numbers listed above.

> **Worked example**
> As an nth term formula you could define $u_n = n^2$. Hence $u_1 = 1^2 = 1$, $u_2 = 2^2 = 4$, $u_3 = 3^2 = 9$ etc., while as an inductive definition you could define:
> $$u_1 = 1 \text{ and } u_n = u_{n-1} + 2n - 1$$
> Hence $u_2 = u_1 + 2(2) - 1 = 1 + 3 = 4$, $u_3 = u_2 + 2(3) - 1 = 4 + 5 = 9$ etc.

Exercise 1

Work out the first five terms of each of these sequences.
(1) $u_n = 3n + 5$
(2) $u_n = 3(2)^n$
(3) $u_1 = 1, u_n = u_{n-1} + n^2$
(4) $u_1 = 1, u_2 = 1, u_n = u_{n-1} + u_{n-2}$

Limits

Consider the sequence defined inductively as $u_1 = 1$ and $u_n = \dfrac{u_{n-1}}{2} + 1$.

An efficient way to investigate this sequence is by using the **ANSWER** button on your calculator (usually **2ND FN and ENTER**). Press '1' (first term) and press 'ENTER'. Then press 'ANSWER $\div$ 2 + 1'. If you now repeatedly press ENTER you get 1, 1.5, 1.75, 1.875,

1.9375, 1.968 75, 1.984 375, 1.992 187 5,.... . These are the terms in the sequence. They are increasing slowly and appear to be getting close to 2. In fact it is said that the sequence **converges** to a **limit** of 2 as **n tends to infinity**. The limiting value is essentially a boundary that cannot be exceeded. You can prove that the limit of the sequence above is 2 using the following argument.

> **Worked example**
>
> Consider $u_1 = 1$ and $u_n = \dfrac{u_{n-1}}{2} + 1$. For a limit to exist, then for some value of n, $u_n = u_{n-1}$ = a fixed value, say L. Substituting L into the inductive formula gives
>
> $L = \dfrac{L}{2} + 1$. Solving this algebraically gives $L = 2$.

So in general, *if you know that a sequence has a limiting value*, you can find this value by substituting L for all the terms with suffixes and then solve to find the value of L.

Exercise 2

Find the limit of the following sequences as n tends to infinity.

(1) $u_1 = 2$, $u_n = \dfrac{u_{n-1}}{3} + 3$

(2) $u_1 = 1$, $u_n = 7 - \dfrac{3u_{n-1}}{4}$

(3) $u_1 = 3$, $u_n = \dfrac{1 + 2u_{n-1}}{8}$

(4) What happens if you try to use the method with the sequence $u_n = 2u_{n-1} + 1$ where $u_1 = 1$? Why is it not valid?

Arithmetic sequences

Any sequence that increases or decreases by a constant amount between successive terms is called an arithmetic sequence. For example, 2, 4, 6, 8 (increases by 2 between each term); 10, 7, 4, 1 (decreases by 3 between each term). The constant amount is called the **common difference**. It is common practice to denote the first term as 'a' and the common difference as 'd'.

Thus, in general, you can define an arithmetic sequence as $u_n = a + (n - 1)d$ (nth term) or $u_n = u_{n-1} + d$, with $u_1 = a$ (inductive).

> **Worked example**
>
> The fifth term of an arithmetic sequence is 16 and the tenth term is 34. Find the first term and the common difference.
>
> *Solution*
>
> Using the formula for the nth term, you know $u_5 = a + (5 - 1)d = a + 4d$, hence $a + 4d = 16$ (equation 1).

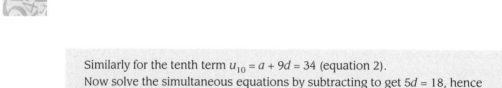

Similarly for the tenth term $u_{10} = a + 9d = 34$ (equation 2).
Now solve the simultaneous equations by subtracting to get $5d = 18$, hence
$d = 3.6$ and $a = 1.6$.

Exercise 3

(1) The seventh term of an arithmetic sequence is 50 and the twelfth term is 75. Find the common difference and the first term.

(2) The first term of an arithmetic sequence is 14 and the common difference is 16. Which term first exceeds 1000?

(3) The tenth term of an arithmetic sequence is 5 and the sixth term is twice the third term. Show that the common difference is 0.5 and find the first term.

Arithmetic series

In order to sum the terms of an arithmetic series efficiently and to solve problems you must generate two formulae. The first one uses a trick first used by a famous German mathematician, Gauss.

Using a = first term, d = common difference, n = number of terms and l = last term, then the terms would be:

$$a, a + d, a + 2d, a + 3d, ..., l - 2d, l - d, l$$

Let S_n denote the sum of these n terms. Hence:

$$S_n = a + (a + d) + (a + 2d) + (a + 3d) + ... + (l - 2d) + (l - d) + l$$

and writing this in reverse gives:

$$S_n = l + (l - d) + (l - 2d) + (a - 3d) + ... + (a + 2d) + (a + d) + a$$

Thus adding each pair of terms 'vertically' gives:

$$2S_n = (a + l) + (a + l) + (a + l) + ... + (a + l) + (a + l)$$

So each of the n brackets is $(a + l)$.

Hence $2S_n = n(a + l)$, therefore $S_n = \frac{n}{2}(a + l)$.

This formula can be used if you know three of the four values: sum (S_n), first term (a), last term (l) and number of terms (n).

By observing that $l = a + (n - 1)d$ (nth term formula) then a second formula is:

$$S_n = \frac{n}{2}[2a + (n - 1)d]$$

This formula can be used if you know three of the four values: sum (S_n), first term (a), common difference (d) and number of terms (n).

Note that you can use these to obtain a formula for the sum of the first n natural numbers: $\frac{n}{2}[n + 1]$.

Sigma notation

Sigma is a Greek capital letter: Σ. Sigma notation represents adding terms together, so it is a convenient way to denote series. It looks like this:

$$\sum_{n=a}^{b} u_n$$

where a = lowest value of the variable, b = highest value of the variable, u_n = formula for sequence. Note that a and b can be thought of as position numbers in the sequence.

Worked example 1

The nth term of an arithmetic sequence is denoted by $u_r = 25 - 3r$.

Evaluate $\sum_{r=1}^{40} u_r$.

Solution

First check the terms: $u_1 = 25 - 3 = 22$, $u_2 = 25 - 6 = 19$, $u_3 = 16,...$
Hence $a = 22$, $d = -3$ and $n = 40$ (from $r = 1$ to $r = 40$ is 40 terms).

Using the formula $S_n = \frac{n}{2}[2a + (n - 1)d]$ gives $S_{40} = \frac{40}{2}[2(22) + (39)(-3)] = -1460$.

Worked example 2

An arithmetic series has first term 2 and common difference 8. The sum to n terms is 3306. Find n.

Solution

Using $S_n = \frac{n}{2}[2a + (n - 1)d]$ with $a = 2$, $d = 8$ and $S_n = 3306$ gives

$\frac{n}{2}[2(2) + (n - 1)(8)] = 3306$.

Hence $\frac{n}{2}[8n - 4] = 3306$. Expanding the bracket gives $4n^2 - 2n = 3306$,
cancelling and rearranging gives $2n^2 - n - 1653 = 0$.
Solving (by formula or factorising) gives $n = -28.5$ or 29. Clearly $n = 29$ (positive whole number).

Worked example 3

Find the sum of all the odd numbers from 501 to 899 inclusive.

Solution

Here $a = 501$, $d = 2$ and $l = 899$. Use $a + (n - 1)d$ to calculate the number of terms.

$$501 + (n-1)2 = 899, \text{ hence } n = \frac{899 - 501}{2} + 1 = 200.$$

$$\text{Then } S_n = \frac{n}{2}(a + l) = \frac{200}{2}(501 + 899) = 140\,000.$$

Exercise 4

(1) The sum of the first 21 terms of an arithmetic series is 525. Show that $a + 10d = 25$. The sum of the first 40 terms of the same arithmetic series is 1760. Hence obtain a second equation and solve both equations to show that the first term is 5 and find the common difference.

(2) Evaluate $\displaystyle\sum_{r=102}^{167} (3r - 1)$.

(3) There are 200 terms in the series $3 + 14 + 25 + \ldots + 2192$. Find the sum of the last 150 terms.

(4) An arithmetic series is such that $u_n = 65 - 4n$. Find the value of m such that

$$\sum_{n=1}^{m} u_n = -882.$$

Geometric sequences

Any sequence that increases or decreases by a constant multiplier between successive terms is called a geometric sequence. For example, 2, 4, 8, 16 (multiplier between each term is 2); 48, 12, 3, 0.75 (multiplier between each term is $\frac{1}{3}$). The constant multiplier is called the **common ratio**. It is common practice to denote the first term as 'a' and the common ratio as 'r'.

Thus in general a geometric sequence is defined as $u_n = ar^{n-1}$ (nth term) or $u_n = ru_{n-1}$, with $u_1 = a$ (inductive).

Worked example

The third term of a geometric sequence is 48 and the sixth term is 3.072. Find the first term and the common ratio.

Solution
Using the formula for the nth term, you know $u_3 = ar^{n-1} = ar^2$, hence $ar^2 = 48$ (equation 1).
Similarly for the sixth term $u_6 = ar^5 = 3.072$ (equation 2).
Now solve the simultaneous equations by dividing to get $r^3 = \dfrac{3.072}{48} = 0.064$ hence $r = 0.4$ and $a = 300$.

Exercise 5

(1) The seventh term of a geometric sequence is 512 and the eleventh term is 2592. Find the common ratio and show that the first term is approximately 45.

(2) The first term of a geometric sequence is 40 and the common ratio is 2. Which is the first term to exceed 5000?

(3) The fourth term of a geometric sequence is 25 times the sixth term. Find the common ratio.

Geometric series

As with arithmetic series it is possible to generate formulae to sum a geometric series.

Using a = first term, r = common ratio, n = number of terms, then the terms would be:

$a, ar, ar^2, ar^3, \ldots , ar^{n-2}, ar^{n-1}$

Let S_n denote the sum of these n terms. Hence:

$S_n = a + ar + ar^2 + ar^3 + \ldots + ar^{n-2} + ar^{n-1}$

Multiplying this by r gives:

$rS_n = ar + ar^2 + ar^3 + \ldots + ar^{n-2} + ar^{n-1} + ar^n$

Then subtracting these equations gives:

$(r - 1)S_n = ar^n - a$ and hence

$$S_n = \frac{ar^n - a}{r - 1} = \frac{a(r^n - 1)}{r - 1} \ or \ \frac{a(1 - r^n)}{1 - r}$$

As a special case consider when $-1 < r < 1$, then as n gets bigger (tends to infinity) r^n will get closer to zero, hence the second formula above simplifies to:

$$\frac{a}{1 - r}$$

This is called the **sum to infinity** of a geometric series:

$$S_\infty = \frac{a}{1 - r}$$

Remember that sums to infinity exist if and only if the common ratio is between -1 and 1. The value of the sum to infinity is essentially a barrier that cannot be exceeded — adding more and more terms gets the sum closer to that value.

Worked example 1

The nth term of a geometric sequence is given by $u_n = 4 \times 3^n$. Write down the first three terms and the common ratio and hence evaluate $\sum_{3}^{15} u_n$, leaving your answer in index form.

Solution

Substituting $n = 1, 2, 3$ gives $u_1 = 12, u_2 = 36, u_3 = 108$. Ratio = 3.

For the stated sum there are 13 terms (15 − 2) so $n = 13$. Since it starts with the value 3, then $a = 108$.

Hence $\displaystyle\sum_{3}^{15} u_n = \frac{a(r^n - 1)}{r - 1} = \frac{108(3^{13} - 1)}{3 - 1} = 54(3^{13} - 1)$.

Worked example 2

The sum to infinity of a geometric series is four times the first term of the series. Find the common ratio, r.

Solution

The first statement means that $\dfrac{a}{1 - r} = 4a$, hence $a = 4a(1 - r)$.

Cancelling and expanding gives $1 = 4 - 4r$.

Hence $4r = 3$ so $r = 0.75$.

Worked example 3

Evaluate $\displaystyle\sum_{n=1}^{\infty} 10 \times 2^{-n}$.

Solution

This requires the sum to infinity. Note that $u_1 = 5$, $u_2 = 2.5$, $u_3 = 1.25$, so $a = 5$ and $r = 0.5$.

Hence using gives $S_\infty = \dfrac{a}{1 - r}$ gives $\dfrac{5}{1 - 0.5} = 10$.

Exercise 6

(1) The nth term of a geometric sequence is $u_n = 5 \times 2^n$. Write down the first three terms and hence find the sum of the first ten terms.

(2) The first term of a geometric sequence is 8 and the fourth term is 343. Find r, the common ratio. Hence determine the lowest value of m such that $S_m > 100\,000$.

(3) Evaluate $\displaystyle\sum_{n=1}^{\infty} 48 \times 3^{-n}$.

(4) The sum to infinity of a geometric series is six times the first term. Find r, the common ratio.

Answers to exercises: types of sequences and series

Exercise 1

(1) 8, 11, 14, 17, 20

(2) 6, 12, 24, 48, 96

(3) 1, 5, 14, 30, 55

(4) 1, 1, 2, 3, 5

Exercise 2

(1) 4.5

(2) 4

(3) $\dfrac{1}{6}$

(4) Sequence is clearly full of positive terms 1, 3, 7, 15,... but limit method gives $L = -1$. Not valid since sequence does not have a limit — it diverges. Method to find the limit is therefore invalid.

Exercise 3

(1) $a = 20, d = 5$

(2) 63rd term (hint: solve $14 + 16(n - 1) > 1000$)

(3) $a = d = 0.5$

Exercise 4

(1) First term = 5, common difference = 2

(2) 26 565 (66 terms)

(3) 205 875

(4) $m = 42$

Exercise 5

(1) $r = 1.5, a = 44.9$ (3 significant figures)

(2) $n = 8$

(3) $r = 0.2$

Exercise 6

(1) $u_1 = 10, u_2 = 20, u_3 = 40, S_{10} = 10\,230$

(2) $r = 3.5, m = 9$

(3) $a = 16, r = \dfrac{1}{3}, S_\circ = 24$

(4) $r = \dfrac{5}{6}$

Trigonometry

Using formulae

Key learning objectives

- to understand and use radian measure — including conversion
- to use sine and cosine rules to solve problems
- to calculate the area of a triangle using the formula $\dfrac{1}{2}ab \sin C$
- to apply the formulae $s = r\theta$ and $A = \dfrac{1}{2}r^2\theta$ to sectors to find lengths, areas or angles

Radians

As an alternative to measuring an angle in degrees you can use **circular measure** where the units are in radians. The use of radians is crucial in calculus work in particular. A simple way to remember radians is to think of a circle with unit radius. The circumference of such a circle would be 2π. So an angle of $360°$ can be represented by 2π radians. Hence:

$$1° = \frac{\pi}{180} \text{ radians}$$

When dealing with angles in radians you can use decimals or for 'convenient' angles you may give them in terms of π. For example, $90° = \frac{\pi}{2}$, $60° = \frac{\pi}{3}$ etc. Scientific calculators have a radian mode shown by the word RAD on the screen or in the option menu. Make sure your calculator is in the correct mode for the examination or that you know how to change it. You can use a ^{c} to denote an angle measured in radians e.g. $2^c = 2$ radians, although it is more common to leave the notation off.

Exercise 1
(1) Express these angles in terms of π.
    (a) 45°
    (b) 30°
    (c) 120°
(2) Convert these angles into degrees, correct to 3 s.f. where appropriate.
    (a) 1 radian
    (b) $\frac{5\pi}{6}$ radians
    (c) 3.5 radians

Solving a triangle

This means finding unknown lengths and angles in any type of triangle. Two main formulae can be used, which were both covered in GCSE Higher Level Mathematics. The correct formula to use depends on the information that has been given, see below. It is convention to denote angles with capital letters A, B, C etc. and the sides opposite these letters with lower case letters a, b, c etc.

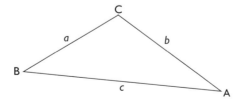

The sine rule is $\dfrac{a}{\sin A} = \dfrac{b}{\sin B} = \dfrac{c}{\sin C}$.

The cosine rule is $a^2 = b^2 + c^2 - 2bc \cos A$ or $\cos A = \dfrac{b^2 + c^2 - a^2}{2bc}$.

For the sine rule to be of use you must know a 'pair' — one angle and the opposite side (same letters). It is used to find:

- a side when you know at least two angles and one side (AAS)
- an angle when you know two sides and an opposite angle (SSA). In this case it is easier to invert the formula so that the sines are on the top

The cosine rule can be used to find:

- a side when you know at least two sides and the angle between them (SAS)
- an angle when you know all three sides. In this case use the rearranged version of the formula

Remember that the three angles in a triangle add up to 180° — so being given two angles means that you know the third.

Another useful formula uses trigonometry to find the area of a triangle. This is usually given as:

$$\text{area} = \frac{1}{2}\, ab \sin C$$

The letters are interchangeable. You must again have the SAS arrangement for this to be used. This formula is simply using the common $\frac{1}{2}$ base × height formula but with the height expressed using sine.

Worked example 1

Find the length of the side BC in the triangle shown.
Calculate the area of the triangle.

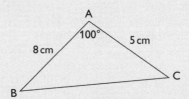

Solution

This is the SAS type so the cosine rule must be used.
The required values are $b = 5$, $c = 8$ and A = 100°.
Using $a^2 = b^2 + c^2 - 2bc \cos A$ gives $a^2 = 5^2 + 8^2 - 2(5)(8)\cos 100° = 102.89...$
Hence $a = \sqrt{102.89} = 10.1$ cm.
Using $\frac{1}{2}\, ab \sin C$ for the area gives area = $\frac{1}{2}$ (5)(8) sin 100° = 19.7 cm^2.

Worked example 2

An acute-angled triangle ABC has AB = 18 cm, AC = 24 cm and angle C = 40°.
Find angle B.

Solution

In this case you have one angle and two sides. The crucial part is that there is an opposite pairing (C = 40° and AB = 18 cm). Hence the sine rule should be used.

The required values are $c = 18$, $b = 24$, $C = 40°$ and B is needed.

Using $\dfrac{a}{\sin A} = \dfrac{b}{\sin B} = \dfrac{c}{\sin C}$ and ignoring the first fraction gives $\dfrac{\sin 40°}{18} = \dfrac{\sin B}{24}$

Hence $\sin B = \dfrac{24 \sin 40°}{18} = 0.857...$ and using inverse sine (sin⁻¹) gives

$B = 58.98...° = 59.0°$.

(Note that because the sine graph repeats itself (see next section) sometimes two different triangles are possible (the ambiguous case); this is why you are told that the triangle is 'acute'.)

Exercise 2

(1) In triangle ABC, AB = 5 cm, BC = 6 cm and AC = 7 cm. Find angle A.

(2) In triangle ABC, AB = 10 cm, BC = 12 cm and angle B = 30°. Find the area of the triangle.

(3) In triangle ABC, AC = 10 cm, angle B = 110° and angle C = 35°. Find BC.

(4) In triangle ABC, AB = 4 cm, BC = 7 cm and angle B = 70°. Find all unknown angles and sides. Is there more than one possibility?

Arc length and area of a sector

Consider the sector shown with radius r and angle θ.

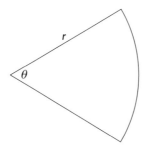

In GCSE Mathematics two formulae were used when θ was in degrees. These were:

arc length = $\dfrac{\theta°}{360°} \times 2\pi r$ and sector area = $\dfrac{\theta°}{360°} \times \pi r^2$

If the angles are measured in radians (remember that $360° = 2\pi$ radians) then the formulae become:

arc length = $r\theta$ and sector area = $\dfrac{1}{2} r^2 \theta$

Questions may require rearrangements of these formulae. A common question is to ask for the area of a segment, as shown below. This requires the difference between the sector area and the triangle area.

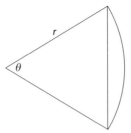

The general formula for the area of the segment is:

$$\frac{1}{2}r^2\theta - \frac{1}{2}r^2\sin\theta$$

Other problems may require the use of the sine rule or cosine rule.

Worked example 1

A sector of a circle has radius 5 cm and angle 1.5 radians. Find the arc length and area of the sector.

Solution

In this case $r = 5$ and $\theta = 1.5$, hence arc length $= 5 \times 1.5 = 7.5$ cm and area $= \frac{1}{2}(5)^2(1.5) = 18.75$ cm^2.

Worked example 2

A sector OAB of a circle has radius 8 cm and a perimeter of 30 cm. The angle between the radii OA and OB is θ radians. Find θ. Hence find the area of the segment bounded by the straight line AB and arc AB.

Solution

Arc length $= 30 - 8 - 8 = 14$ cm.

Using arc length $= r\theta$, you get $14 = 8\theta$ and hence $\theta = 1.75$ radians.

Area of sector $= \frac{1}{2}r^2\theta = \frac{1}{2}(8)^2(1.75) = 56$ cm^2

Area of triangle $= \frac{1}{2}r^2\sin\theta = \frac{1}{2}(8)^2\sin 1.75 = 31.48....$ cm^2

Hence segment area $= 56 - 31.48 = 24.5$ cm^2.

Exercise 3

(1) A sector of a circle has radius 10 cm and angle 2.5 radians. Find the arc length and area of the sector.

(2) A sector of a circle with radius 4 cm and angle θ radians has an area of 6 cm^2. Find θ and determine the perimeter of the sector.

(3) OAB is a sector of a circle with centre O and radius r cm. The angle AOB = 1.2 radians. The perimeter of the sector is 25.6 cm. Find r and determine the area of the segment bounded by the straight line AB and the arc AB.

(4) A sector of a circle radius r and angle θ is such that the magnitude of the area equals the magnitude of the arc length. Find r.

Answers to exercises: using formulae

Exercise 1

(1) (a) $\frac{\pi}{4}$

(b) $\frac{\pi}{6}$

(c) $\frac{2\pi}{3}$

(2) (a) $57.3°$

(b) $150°$

(c) $201°$

Exercise 2

(1) $57.1°$

(2) $30\,\text{cm}^2$

(3) $6.10\,\text{cm}$

(4) AC = 6.77 cm, angle A = 76.3°, angle C = 33.7°; or AC = 6.77 cm, angle A = 103.7°, angle C = 6.3°

Exercise 3

(1) Arc length = 25 cm, sector area = 125 cm^2

(2) $\theta = 0.75$, perimeter = 11 cm

(3) $r = 8$, segment area = 8.57 cm^2

(4) $r = 2$

Solving equations and proving identities

Key learning objectives

- to sketch the standard graphs of sine, cosine and tangent functions
- to understand and appreciate key properties of the standard graphs
- to solve trigonometrical equations of the form $a\sin(bx + c) = k$, $a\cos(bx + c) = k$ or $a\tan(bx + c) = k$ within a given range
- to learn and apply the identities $\tan\theta = \frac{\sin\theta}{\cos\theta}$ and $\sin^2\theta + \cos^2\theta = 1$ to a range of equations and expressions
- to solve quadratic equations in sine, cosine or tangent

Graphs of trigonometrical functions

The standard trigonometrical graphs of sine, cosine and tangent are shown below for x between 0° and 360°. All three graphs are **periodic** — they repeat after a fixed period. This varies depending on the function involved. The properties of each graph are given below. n represents an integer.

Graph of y = sin x

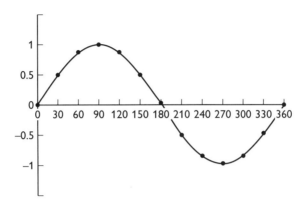

- Graph has a period of 360°.
- Highest value = 1 when $x = 90°$ (in general $x = 360°n + 90°$).
- Lowest value of –1 when $x = 270°$ (in general $x = 360°n - 90°$).
- Rotational symmetry of order 2 about the origin (in general about $(180°n, 0)$).
- Lines of reflection symmetry at $x = 180°n - 90°$.
- Sine is an odd function.

Graph of y = cos x

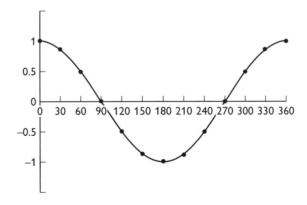

- Graph has a period of 360°.
- Highest value = 1 when $x = 0°$, 360° (in general $x = 360°n$).
- Lowest value of –1 when $x = 180°$ (in general $x = 360°n - 180°$).
- Rotational symmetry of order 2 about $(180°n - 90°, 0)$.
- Reflection symmetry about the y-axis (in general about $x = 180°n$).
- Cosine is an even function.

Graph of y = tan x

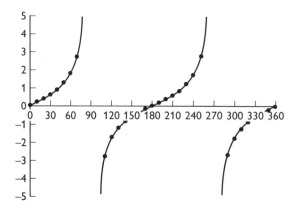

- Graph has a period of 180°.
- No highest or lowest value.
- Asymptotes (where curve does not exist) when $x = 90°$ and 270° (in general $x = 180°n + 90°$).
- Rotational symmetry of order 2 about (0, 0) — in general about (180°n , 0).
- Tangent is an odd function.

All other sine, cosine and tangent graphs are linked to the above three. See transformations on p. 17.

Solving simple trigonometrical equations

To solve any simple trigonometrical equation, use the inverse trigonometrical functions on your calculator and knowledge of the graph properties above. Since all the trigonometrical graphs are periodic and repeat, a solution range will be stated in the question. When using the inverse trigonometrical functions on your calculator, only one value will be given, this value is called the **principal value**, although the **calculator value** is equally appropriate. Details of these are given below as they vary from function to function:

- For the sine curve the principal value of x is such that $-90° \le x \le 90°$.
- For the cosine curve the principal value of x is such that $0° \le x \le 180°$.
- For the tangent curve the principal value of x is such that $-90° < x < 90°$.

In the graphs above, the x-axis is in degrees. It could be in radians, in which case it is usual to mark key values in terms of π (remember that $180° = \pi$ radians). The respective periods for sine and cosine are 2π radians, while for tangent it is π radians. The principal values are adjusted in a similar manner.

The algorithm for solving simple trigonometrical equations is as follows.

(1) The key idea is to make sure that the equation is in the form:
      TRIG FUNCTION(VARIABLE) = NUMBER.

(2) Use the appropriate inverse trigonometrical function to obtain the principal value (calculator value) of the variable. Make sure your calculator is in the correct mode, depending on whether the angle is required in degrees or radians.

(3) Use the symmetries and/or periodicity of the appropriate graph to obtain the other solutions in the required range.

Worked example 1

Solve the equation $\sin \theta = 0.5$ for $0° \leq \theta \leq 720°$.

Solution

First find the principal value of $\theta = \sin^{-1} 0.5 = 30°$.

On the standard sine graph, it can be seen that $150°$ (note $150° = 180° - 30°$) is the same height as $30°$ so these are the first two solutions.

Then use the fact that the graph is periodic and add $360°$ to each of these solutions to get $30° + 360° = 390°$ and $150° + 360° = 510°$. If you try to add $360°$ again, the values are outside the range stated.

There are therefore four solutions: $\theta = 30°$, $150°$, $390°$ and $510°$.

Worked example 2

Solve the equation $2\tan x + 4 = 0$ for $0 \leq x \leq 2\pi$.

Solution

Note that the solutions are required in radians.

First rearrange to get the correct format of $\tan x = -2$.

Now find the principal value of $\tan^{-1} (-2) = -1.107$ radians.

This is outside the range so add on π (period for tan in radians) to get $-1.107 + \pi = 2.034$.

Then add another π to get $2.034 + \pi = 5.176$. If another π is added then the value is outside the given range ($2\pi = 6.283$). Therefore there are two solutions: $x = 2.034$ and 5.176.

Worked example 3

Solve the equation $\sqrt{2} = 2\cos \theta$ for the range $-180° \leq \theta \leq 180°$.

Solution

Rearranging gives $\cos \theta = \dfrac{\sqrt{2}}{2}$.

The principal value is given by $\cos^{-1} \left(\dfrac{\sqrt{2}}{2} \right) = 45°$.

From the standard graph it can be seen that $315°$ ($360° - 45°$) is the same height as $45°$; this would be the second solution but is outside the range. On this occasion we must subtract $360°$ (period of the cos graph) to generate solutions in the correct range.

$45° - 360° = -315°$ — out of range.

$315° - 360° = -45°$ — in range. Hence solutions are: $\theta = -45°$ and $45°$.

Remembering the general structure from each solution can help to find all the specific solutions. The general solutions are used in the module Further Pure 1 but can also help here. In the formulae listed below, n is an integer and PV = principal value (calculator value).

- For a tangent graph: **variable** = PV + 180°n.
- For a cosine graph: **variable** = ± PV + 360°n.
- For a sine graph: **variable** = PV + 360°n and variable = (180° – PV) + 360°n.

(Note that when working in radians, replace the 180° by π and 360° by 2π.)

Harder equations will include simple functions of the variable such as 3θ or $\theta - 20°$ or $\theta + 50°$.

To find all the solutions in these cases the solution range must first be adjusted. The next two examples illustrate this.

Worked example 1

Solve $\sin 3\theta = 0.9$ for $0 \le \theta \le 180°$ giving your answers to 1 decimal place.

Solution

To work out the adjusted range use 3θ, so $0 \le 3\theta \le 540°$, so initial solutions for 3θ must be generated up to and including 540°.
Principal value = $\sin^{-1} 0.9 = 64.16°$. From the general graph (or formula) the other solution would be $180° - 64.16° = 115.84°$.
Adding on multiples of 360° to each of these in turn gives $64.16° + 360° = 424.16°$ and $115.84° + 360° = 475.84°$. All other values are outside the range $0 \le 3\theta \le 540°$.
Hence $3\theta = 64.16°, 115.84°, 424.16°$ and $475.84°$. (Note that 3θ *must* be used at this point.)
Hence, dividing by 3 gives $\theta = 21.4°, 38.6°, 141.4°$ and $158.6°$.
The next example uses one of the general structure formulae to generate solutions.

Worked example 2

Solve the equation $\cos (x - \frac{\pi}{6}) = -\frac{1}{2}$ for $-2\pi \le x \le 2\pi$, giving answers in terms of π.

Solution

Adjust the range by subtracting $\frac{\pi}{6}$ to get $\frac{-13}{6}\pi \le x - \frac{\pi}{6} \le \frac{11}{6}\pi$.

$\cos^{-1}(-\frac{1}{2}) = \frac{2\pi}{3}$ so the PV = $\frac{2\pi}{3}$ and using the general structure formula gives

$x - \frac{\pi}{6} = \pm\frac{2\pi}{3} + 2\pi n$ so substituting n values ($n = 0, \pm1, \pm2$ etc.) gives

$x - \frac{\pi}{6} = -\frac{4\pi}{3}, -\frac{2\pi}{3}, \frac{2\pi}{3}, \frac{4\pi}{3}$ (no others are in the range).

Adding $\frac{\pi}{6}$ to each of these gives $x = -\frac{7\pi}{6}, -\frac{\pi}{2}, \frac{5\pi}{6}, \frac{3\pi}{2}$.

Exercise 1

Solve each of the following for the range stated:

(a) $3\sin x - 1 = 0$ for $0 \le x \le 2\pi$

(b) $\tan x = 1$ for $0 \le x \le 720°$

(c) $\cos 2\theta = -0.3$ for $-\pi \le \theta \le \pi$

(d) $2\sin(\theta + 20°) - 1 = 0$ for $-180° \le \theta \le 540°$

Solving quadratic equations

If an equation has the structure $a[f(x)]^2 + b[f(x)] + c = 0$ where $f(x) = \sin kx, \cos kx$ **or $\tan kx$**, then the equation is a quadratic in that trigonometrical function. You can therefore factorise, complete the square or use the quadratic formula to reduce the quadratic equation to two simple trigonometrical equations and solve each using the method described above. Sometimes one of these equations may yield no valid solutions for the trigonometrical function involved — always state this clearly if it occurs.

> **Worked example**
>
> Solve $2\cos^2\theta + 3\cos\theta - 2 = 0$ for $0 \le \theta \le 360°$.
>
> **Solution**
>
> Let $c = \cos\theta$ so $2c^2 + 3c - 2 = 0$.
>
> This factorises into $(2c - 1)(c + 2) = 0$ and hence $c = \frac{1}{2}$ or $c = -2$.
>
> Replacing c with $\cos\theta$ gives $\cos\theta = \frac{1}{2}$ or $\cos\theta = -2$.
>
> The second of these is impossible (see the standard graph), hence solving $\cos\theta = \frac{1}{2}$ using the methods described earlier, the solutions are $60°$ and $300°$.

Exercise 2

Solve each of these equations in the stated range:

(a) $2\sin^2\theta - 3\sin\theta + 1 = 0$ for $0 \le \theta \le 360°$

(b) $\tan^2\theta + \tan\theta - 2 = 0$ for $0 \le \theta \le 360°$

(c) $\cos^2\theta - 1 = 0$ for $0 \le \theta \le 2\pi$

Important identities

Consider this right-angled triangle with hypotenuse 1 unit:

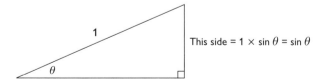

This side = $1 \times \sin\theta = \sin\theta$

This side = $1 \times \cos\theta = \cos\theta$

Using the definition of tangent gives $\tan\theta = \dfrac{\sin\theta}{\cos\theta}$ **(Identity 1)** and using Pythagoras' rule gives $\sin^2\theta + \cos^2\theta = 1$ **(Identity 2)**.

Both these identities are important because they enable us to prove identities and solve other types of equations. Identity 2 helps to convert expressions or equations in terms of sine or cosine only. This usually results in a quadratic equation which can then be solved as described above.

Worked example

Show that $2 \tan \theta \cos^3 \theta = 2 \sin \theta (1 - \sin \theta)(1 + \sin \theta)$.

Solution

Start by replacing $\tan \theta = \dfrac{\sin \theta}{\cos \theta}$ to get $2 \dfrac{\sin \theta}{\cos \theta} \cos^3 \theta$. Then cancel $\cos \theta$ to get $2 \sin \theta \cos^2 \theta$.

From **Identity 2**, $\cos^2 \theta = 1 - \sin^2 \theta$, so $2 \sin \theta \cos^2 \theta = 2 \sin \theta (1 - \sin^2 \theta)$ and factorising gives $2 \sin \theta (1 - \sin \theta)(1 + \sin \theta)$.

Exercise 3

(1) Express $2\sin^2 \theta + \cos^2 \theta$ in terms of cosine only.

(2) Express $3\cos^2 \theta - \sin^2 \theta - 3$ in terms of sine only.

(3) Express $2\tan \theta \sin \theta \cos \theta$ in terms of cosine only.

Worked example 1

Solve $\sin x - 2\cos x = 0$ for $0 \leq x \leq 2\pi$.

Solution

First rewrite as $\sin x = 2\cos x$.

Dividing by $\cos x$ gives $\dfrac{\sin x}{\cos x} = 2$

Dividing by $\cos x$ is allowed because if $\cos x = 0$, $x = \pm \dfrac{\pi}{2}$ and $\sin \left(\pm \dfrac{\pi}{2} \right) - 2 \cos \left(\pm \dfrac{\pi}{2} \right) \neq 0$.

Therefore using the identity, $\tan x = 2$.

Solving this using the method for solving simple trigonometric equations gives $x = 1.11$ and 4.25.

Worked example 2

Solve $3\cos^2 \theta + 4\sin \theta = 4$ for $0 \leq x \leq 360°$.

Solution

Using $\cos^2 \theta = 1 - \sin^2 \theta$ and substituting gives $3(1 - \sin^2 \theta) + 4\sin \theta = 4$.

Expanding the bracket gives $3 - 3\sin^2 \theta + 4\sin \theta = 4$.

Collecting all terms on one side gives $3\sin^2 \theta - 4\sin \theta + 1 = 0$.

This factorises into $(3\sin \theta - 1)(\sin \theta - 1) = 0$.

Hence either $\sin \theta = \dfrac{1}{3}$ or $\sin \theta = 1$.

Solving these gives $\theta = 19.5°$, $160.5°$ (first equation) and $\theta = 90°$ (second equation).

Exercise 4

(1) Solve $2\sin x + \cos x = 0$ for $0 \leq x \leq 360°$.

(2) Solve $\sin^2 \theta = 3(\cos \theta + 1)$ for $0 \leq x \leq 360°$.

(3) Solve $2\cos^2 \theta + \sin \theta = 1$ for $0 \leq \theta \leq 2\pi$.

Answers to exercises: solving equations and proving identities

Exercise 1

(a) 0.340, 2.80

(b) 45°, 225°, 405°, 585°

(c) −2.20, −0.938, 0.938, 2.20

(d) 10°, 130°, 370°, 490°

Exercise 2

(a) 30°, 90°, 150°

(b) 45°, 117°, 225°, 297°

(c) 0, π, 2π or 0, 3.14, 6.28

Exercise 3

(1) $2 - \cos^2 \theta$

(2) $-4\sin^2 \theta$

(3) $2(1 - \cos^2 \theta)$

Exercise 4

(1) 153°, 333°

(2) 180°

(3) 1.57, 3.67, 5.76 or $\dfrac{\pi}{2}, \dfrac{7\pi}{6}, \dfrac{11\pi}{6}$

Exponentials and logarithms

Key learning objectives

- to be aware of the standard graph of $y = a^x$ and its properties
- to appreciate the link between exponentials and logarithms, being able to convert between them
- to learn and apply the three laws of logarithms, in particular to simplify expressions
- to solve simple exponential equations by taking logs
- to be able to convert between bases

Graph of $y = a^x$

Functions such as $y = 2^x$ and $y = 3^x$ are called exponential functions. The general function is $y = a^x$. For $a > 1$, the curve is shown on p. 48.

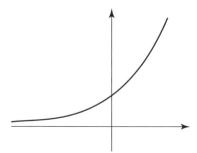

- It crosses the y-axis at (0, 1).
- It increases rapidly as x increases.
- It approaches 0 as x approaches large negative values.
- It is an increasing function.

If $0 < a < 1$, the curve looks like this:

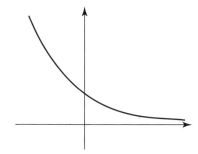

- It also crosses the y-axis at (0, 1).
- It decreases rapidly as x increases.
- It approaches 0 as x approaches large positive values.
- It is a decreasing function.

It is important to know the properties listed above and to be able to transform such graphs using the ideas from the first section — see 'Functions' (p. 17). Questions may require sketches.

Powers, bases and logarithms

It is often useful to express a number y in power form, so that $y = a^x$. The 'x' value is called a **logarithm** and the 'a' value is called the **base**. This is said as '**the logarithm of y to base a is x**' and is written as $\log_a y = x$. It follows that for every power statement there is a corresponding logarithm statement. For example:

$10^2 = 100$ therefore $\log_{10} 100 = 2$

$4^3 = 64$ therefore $\log_4 64 = 3$

$2^{-1} = \frac{1}{2}$ therefore $\log_2 \frac{1}{2} = -1$

Two important results are $\log_a a = 1$ (since $a^1 = a$) and $\log_a 1 = 0$ (since $a^0 = 1$).

Exercise 1

By thinking of the appropriate power statement, evaluate x.

(a) $x = \log_3 81$

(b) $x = \log_2 64$

(c) $x = \log_6 \dfrac{1}{36}$

(d) $\log_2 x = 4$

(e) $\log_5 x = -1$

(f) $\log_{100} x = \dfrac{1}{2}$

On a scientific calculator, logarithms to base 10 can be easily evaluated. Simply press the 'log' button and enter the number. Other bases can be evaluated by using the formula for the change of base. This formula is:

$$\log_a b = \frac{\log_c b}{\log_c a}$$

so in particular $\log_a b = \dfrac{\log_{10} 7}{\log_{10} 2} = \dfrac{\log_{10}(\text{original number})}{\log_a(\text{original base number})}$

For example $\log_2 7 = \dfrac{\log_{10} 7}{\log_{10} 2} = 2.81$ (3 significant figures). The '10' is not usually included as 'log' refers to base 10.

Exercise 2

Evaluate the following correct to 3 significant figures:

(a) $\log_3 24$

(b) $\log_2 0.4$

(c) $\log_4 100$

Laws of logarithms

There are three main laws of logarithms — related to some of the laws of indices.

These laws are:

- $\log_a (mn) = \log_a m + \log_a n$ (Law 1: multiplying numbers — add the logs)

- $\log_a \left(\dfrac{m}{m}\right) = \log_a m - \log_a n$ (Law 2: dividing numbers — subtract the logs)

- $\log_a x^n = n\log_a x$ (Law 3: repeated multiplication — bring the power down)

The main uses of these are to combine several numerical logs to a single log, to simplify algebraic expressions and to solve equations (exponential or logs). The next examples illustrate these ideas.

Worked example 1

Express $3\log_a 2 + \dfrac{1}{2} \log_a 16 - \log_a 1.6$ in the form $\log_a p$, where p is an integer to be found.

Solution

$3\log_a 2 = \log_a 2^3 = \log_a 8$ (using **Law 3**) and $\frac{1}{2}\log_a 16 = \log_a 16^{\frac{1}{2}} = \log_a 4$ (using **Law 3**)

Hence expression becomes $\log_a 8 + \log_a 4 - \log_a 1.6$

Then $\log_a 8 + \log_a 4 = \log_a (8 \times 4) = \log_a 32$ (using **Law 1**)

Hence $\log_a 8 + \log_a 4 - \log_a 1.6 = \log_a 32 - \log_a 1.6 = \log_a \left(\frac{32}{1.6}\right) = \log_a 20$ (using **Law 2**)

Worked example 2

Given that $\log_a x - 4\log_a 3 = 5$, express x in terms of a.

Solution

First observe that $4\log_a 3 = \log_a 3^4 = \log_a 81$ (using **Law 3**)

Then $\log_a x - 4\log_a 3 = \log_a x - \log_a 81 = \log_a \frac{x}{81}$ (using **Law 2**)

Hence $\log_a \frac{x}{81} = 5$, which gives the power statement $\frac{x}{81} = a^5$ and so $x = 81a^5$

Worked example 3

It is given that $\log_2 k = 3 - \log_2 (k - 2)$. Determine the value of k.

Solution

First rewrite to get $\log_2 k + \log_2 (k - 2) = 3$

Then use **Law 1** to get $\log_2 k(k - 2) = 3$

Hence the power statement gives $k(k - 2) = 2^3 = 8$

Then multiplying the bracket gives $k^2 - 2k = 8$, so $k^2 - 2k - 8 = 0$

Factorising gives $(k - 4)(k + 2) = 0$ so $k = 4$ or -2. The latter value cannot apply because $\log_2 (-2)$ does not exist. Hence $k = 4$.

Worked example 4

Solve $4^x = 12$, giving your answer to 3 significant figures.

Solution

Take the log of both sides to get $\log_{10} 4^x = \log_{10} 12$

Use **Law 3** to 'bring the power down' so $x\log_{10} 4 = \log_{10} 12$

Now divide by $\log_{10} 4$ to get $x = \dfrac{\log_{10} 12}{\log_{10} 4} = 1.79$

(Note that rather than writing $\log_{10}$, you can just write log.)

Worked example 5

Solve $3^{2x-1} = 20$, giving your answer to 3 significant figures.

Solution

Take the log of both sides to get $\log 3^{2x-1} = \log 20$

Use **Law 3** to 'bring the power down' so $(2x - 1)\log 3 = \log 20$

Divide by $\log 3$ to get $2x - 1 = \dfrac{\log 20}{\log 3} = 2.727$ (allow an extra figure)

Now solve the linear equation to get $x = \dfrac{1}{2}(2.727 + 1) = 1.86$

Exercise 3

(1) Solve $5^{x+1} = 28$, giving you answer to 3 significant figures.

(2) By writing $y = 2^x$, show that $2^{2x} - 12(2^x) + 32 = 0$ becomes $y^2 - 12y + 32 = 0$. Solve this equation for y and hence deduce the two possible values of x.

(3) Given that $\log_a x = 4\log_a 2 - \dfrac{1}{2}\log_a 4$, find x.

(4) Given that $\log_a x = z + n\log_a y$, find an expression for x in terms of a, y, z and n.

(5) Given that $2\log_a k = \log_a 2 + \log_a (k + 24)$, show that $k^2 - 2k - 48 = 0$ and deduce the value of k.

Answers to exercises: exponentials and logarithms

Exercise 1

(a) 4

(b) 6

(c) –2

(d) 16

(e) $\dfrac{1}{5}$

(f) 10

Exercise 2

(a) 2.89

(b) –1.32

(c) 3.32

Exercise 3

(1) 1.07

(2) $y = 4$ or $y = 8$, $x = 2$ or $x = 3$

(3) 8

(4) $x = a^z y^n$

(5) $k = 8$ (–6 is not possible)

Differentiation

Key learning objectives

- to differentiate terms of the form ax^n where n is any rational number
- to use a range of algebraic techniques to rewrite expressions so that they can be differentiated
- to solve a range of problems involving gradients, tangents, normals, stationary points, optimisation etc. — extending knowledge from Core 1

In Core 1 you learnt that when $y = ax^n$ then $\dfrac{dy}{dx} = anx^{n-1}$. In words the rule was 'multiply by the power and then subtract 1 from the power'. This rule was only used when n was a positive integer. By using knowledge of indices you can now extend this idea to differentiate any term where n is any rational number. The two crucial results are:

$$\frac{1}{x^n} = x^{-n} \text{ and } \sqrt[n]{x} = x^{\frac{1}{n}}.$$

Consider the results shown below for negative powers of x:

y	y rewritten using negative powers	$\dfrac{dy}{dx}$	$\dfrac{dy}{dx}$ rewritten without negative powers
$\dfrac{1}{x}$	x^{-1}	$-x^{-2}$	$-\dfrac{1}{x^2}$
$\dfrac{1}{x^2}$	x^{-2}	$-2x^{-3}$	$-\dfrac{2}{x^3}$
$\dfrac{1}{x^3}$	x^{-3}	$-3x^{-4}$	$-\dfrac{3}{x^4}$
$\dfrac{1}{x^n}$	x^{-n}	$-nx^{-n-1}$	$-\dfrac{n}{x^{n+1}}$

Similarly the next set of results relate to fractional powers of x:

y	y rewritten using fractional powers	$\dfrac{dy}{dx}$	$\dfrac{dy}{dx}$ rewritten without negative or fractional powers
$\sqrt{x}$	$x^{\frac{1}{2}}$	$\dfrac{1}{2}x^{-\frac{1}{2}}$	$\dfrac{1}{2\sqrt{x}}$
$\sqrt[3]{x}$	$x^{\frac{1}{3}}$	$\dfrac{1}{3}x^{-\frac{2}{3}}$	$\dfrac{1}{3\sqrt[3]{x^2}}$
$\sqrt[n]{x}$	$x^{\frac{1}{n}}$	$\dfrac{1}{n}x^{-\left(\frac{n-1}{n}\right)}$	$\dfrac{1}{n\sqrt[n]{x^{n-1}}}$
$\sqrt[n]{x^m}$	$x^{\frac{m}{n}}$	$\dfrac{m}{n}x^{-\left(\frac{n-m}{n}\right)}$	$\dfrac{m}{n\sqrt[n]{x^{n-m}}}$

The following have a mixture of terms that need to be rewritten along with some that do not.

Worked example 1

Differentiate $y = 5x^3 - 4x + \dfrac{6}{x^4}$.

Solution

The third term must be rewritten so $y = 5x^3 - 4x + 6x^{-4}$

Hence $\dfrac{dy}{dx} = 5 \times 3x^{3-1} - 4 \times 1x^{1-1} + 6 \times (-4)x^{-4-1}$

Which simplifies to $\dfrac{dy}{dx} = 15x^2 - 4 - 24x^{-5}$ or $15x^2 - 4 - \dfrac{24}{x^5}$

Worked example 2

Differentiate $y = 2x^2 - 4\sqrt{x}$.

Solution

The second term needs to be rewritten using fractional powers.

So $y = 2x^2 - 4x^{\frac{1}{2}}$

Therefore $\dfrac{dy}{dx} = 2 \times 2x^{2-1} - 4 \times \dfrac{1}{2}\, x^{\frac{1}{2}-1}$

Which simplifies to $\dfrac{dy}{dx} = 4x - 2x^{-\frac{1}{2}}$ or $4x - \dfrac{2}{\sqrt{x}}$

Exercise 1

Rewrite then differentiate each of the following.

(a) $y = 10x^2 - \dfrac{1}{x}$

(b) $y = 6x - \dfrac{3}{x^2}$

(c) $y = 20 - 4\sqrt{x}$

(d) $y = \sqrt[3]{x} - \sqrt[4]{x}$

(e) $y = 7x^3 - \dfrac{4}{\sqrt{x}}$

In addition some questions may require the use of two further rules of indices:

$$x^m x^n = x^{m+n} \text{ and } \dfrac{x^m}{x^n} = x^{m-n}$$

Worked example 1

Express $x^3\sqrt{x}$ in the form x^p where p is a rational number. Hence differentiate $x^3\sqrt{x}$.

Solution

First rewrite as $x^3 x^{\frac{1}{2}}$ then use the above rule to write as a single power so

it becomes $x^{3+\frac{1}{2}} = x^{\frac{7}{2}}$

Differentiating gives $\frac{7}{2} x^{\frac{5}{2}}$ or $\frac{7}{2}\sqrt{x^5}$

Worked example 2

Express $\dfrac{x(\sqrt{x}-1)}{x^2}$ in the form $x^p - x^q$ where p and q are rational numbers.

Hence differentiate $\dfrac{x(\sqrt{x}-1)}{x^2}$ 4.

Solution

First multiply out the bracket to get $\dfrac{x\sqrt{x} - x}{x^2}$

Rewrite the numerator using powers to get $\dfrac{x^{\frac{3}{2}} - x}{x^2}$

Split into two fractions to get $\dfrac{x^{\frac{3}{2}}}{x^2} - \dfrac{x}{x^2}$

Then use the second rule above to get $x^{-\frac{1}{2}} - x^{-1}$ which is now in the required form.

Differentiating gives $-\dfrac{1}{2} x^{-\frac{3}{2}} + x^{-2}$ or $-\dfrac{1}{2\sqrt{x^3}} + \dfrac{1}{x^2}$

Exercise 2

In each case, rewrite in the form x^p where p is a rational number, then differentiate the expression.

(a) $x^2\sqrt{x}$

(b) $x\sqrt[3]{x}$

In each case rewrite in the form $x^p + x^q$ where p and q are rational numbers and hence differentiate the expression.

(c) $\sqrt{x}(x^2 + 1)$

(d) $\dfrac{x^3 + \sqrt{x}}{x^5}$

As in Core 1, questions about differentiation can be linked to specific gradients, equations of tangents and normals, increasing or decreasing functions, stationary points and a variety of optimisation problems. A brief reminder about each one is given below before a number of examples are given and brief exercises follow.

Specific gradients

Differentiate first then substitute the appropriate x value into $\dfrac{dy}{dx}$ to obtain the required gradient. If a gradient is given and the point is required, again differentiate first then put the derivative equal to the given value for the gradient and solve for x. The value of y can be found by substituting the x value into the formula for y.

Equations of tangents and normals

First differentiate and substitute the x value to find the gradient of the tangent (as above). To find the equation of the tangent, use the formula $y - y_1 = m(x - x_1)$ where (x_1, y_1) is the given point and m is the gradient of the tangent. The gradient of the normal will be $-\dfrac{1}{m}$ and therefore the equation is given by $y - y_1 = -\dfrac{1}{m}(x - x_1)$.

Stationary points and optimisation problems

Differentiate, then solve $\dfrac{dy}{dx} = 0$ to find the x values of any stationary points. The y value can be found by using the formula for y — it is this value that is the **maximum or minimum value** of the curve being considered. To determine the nature of the stationary point, differentiate again to find the second derivative $\dfrac{d^2 y}{dx^2}$. Then substitute each x value already obtained.

- If $\dfrac{d^2 y}{dx^2} < 0$ then the stationary point is a maximum.

- If $\dfrac{d^2 y}{dx^2} > 0$ then the stationary point is a minimum.

- In the case when $\dfrac{d^2 y}{dx^2} = 0$ the test is inconclusive and you need to test $\dfrac{dy}{dx}$ either side of the x value to determine the true nature.

A point of inflexion would have a gradient with the same sign either side of the x value used. Optimisation problems relate to practical problems involving area, volume etc. An algebraic expression will need to be formed first before the above technique is used to locate the stationary points and hence the maximum or minimum values.

Increasing or decreasing functions

To prove that a function is always increasing or decreasing, differentiate to find $\dfrac{dy}{dx}$, then attempt to rewrite it so that it can be seen that the value is always greater than 0 (**increasing function**) or less than 0 (**decreasing function**). Any algebraic technique can be used, but often completing the square is required.

Worked example 1

Find the equation of the normal to the curve $y = 4x + \dfrac{1}{\sqrt{x}}$ at the point when $x = 1$.

Solution

First rewrite to get $y = 4x + x^{-\frac{1}{2}}$

Then differentiate to get $\dfrac{dy}{dx} = 4 - \dfrac{1}{2} x^{-\frac{3}{2}}$ or $4 - \dfrac{1}{2\sqrt{x^3}}$

Substitute $x = 1$ to get the gradient of the tangent $= 4 - \dfrac{1}{2\sqrt{1^3}} = \dfrac{7}{2}$

Hence the gradient of the normal $= -\dfrac{2}{7}$

Substitute $x = 1$ into the y formula to get $y = 4(1) + \dfrac{1}{\sqrt{1}} = 5$

Now use $y - y_1 = m(x - x_1)$ to get the equation of the normal as $y - 5 = -\dfrac{2}{7}(x - 1)$

Tidying this up gives $7y + 2x - 37 = 0$

Worked example 2

Find the coordinates of the stationary points on the curve $y = 16x^2 + \dfrac{1}{x^2}$ and determine their nature.

Solution

First rewrite to get $y = 16x^2 + x^{-2}$

Then differentiate to get $\dfrac{dy}{dx} = 32x - 2x^{-3}$ or $32x - \dfrac{2}{x^3}$

Now solve $32x - \dfrac{2}{x^3} = 0$

So $32x - \dfrac{2}{x^3}$ and therefore $x^4 = \dfrac{1}{16}$

So $x = \sqrt[4]{\dfrac{1}{16}} = \dfrac{1}{2}$ or $-\dfrac{1}{2}$

Now find the y values by substituting these x values into the y formula

So when $x = \dfrac{1}{2}$, $y = 16\left(\dfrac{1}{2}\right)^2 + \dfrac{1}{\left(\dfrac{1}{2}\right)^2} = 8$

Similarly when $x = -\dfrac{1}{2}$, $y = 8$

Now differentiate again to get $\dfrac{d^2y}{dx^2} = 32 + 6x^{-4}$ or $32 + \dfrac{6}{x^4}$

Substituting $x = \dfrac{1}{2}$, gives $\dfrac{d^2y}{dx^2} = 128 > 0$ and hence is a minimum point, with minimum value 8.

Similarly for $x = \dfrac{1}{2}$, $\dfrac{d^2y}{dx^2} = 128 > 0$ and hence is also a minimum point, with minimum value 8.

Worked example 3

Show that the curve $y = \dfrac{10}{x} - 3x$, $x \neq 0$ is a decreasing function for all values of x.

Solution

First rewrite to get $y = 10x^{-1} - 3x$

Differentiate to get $\dfrac{dy}{dx} = -\dfrac{10}{x^2} - 3$

x^2 is always positive (since $x \neq 0$) so $-\dfrac{10}{x^2}$ is always negative, subtracting another 3 still leaves it negative.

Hence $\dfrac{dy}{dx}$ is always negative and the curve is therefore decreasing.

Exercise 3

(1) Find the equation of the tangent to the curve $y = \sqrt[3]{x}$ at the point (8, 2).

(2) Find the equation of the normal to the curve $y = x^3 - \dfrac{2}{x^3}$ at the point $x = -1$.

(3) Find the coordinates of the point on the curve $y = \sqrt{x} - 1$ at which the gradient is parallel to the line $8y - x = 0$.

(4) Find the coordinates of the stationary point on the curve $y = x + \dfrac{25}{x}$ and determine their nature.

(5) Show that the curve $y = x^3 - \dfrac{1}{x} - \dfrac{2}{x^3}$, $x \neq 0$ is an increasing function.

Answers to exercises: differentiation

Exercise 1

(a) Rewritten as $10x^2 - x^{-1}$. Derivative is $20x + x^{-2}$ or $20x + \dfrac{1}{x^2}$

(b) Rewritten as $6x - 3x^{-2}$. Derivative is $6 + 6x^{-3}$ or $6 + \dfrac{6}{x^3}$

(c) Rewritten as $20 - 4x^{\frac{1}{2}}$. Derivative is $-2x^{-\frac{1}{2}}$ or $\dfrac{-2}{\sqrt{x}}$

(d) Rewritten as $x^{\frac{1}{3}} - x^{\frac{1}{4}}$. Derivative is $\dfrac{1}{3}x^{-\frac{2}{3}} - \dfrac{1}{4}x^{-\frac{3}{4}}$ or $\dfrac{1}{3\sqrt[3]{x^2}} - \dfrac{1}{4\sqrt[4]{x^3}}$

(e) Rewritten as $7x^3 - 4x^{-\frac{1}{2}}$. Derivative is $21x^2 + 2x^{-\frac{3}{2}}$ or $21x^2 + \dfrac{2}{\sqrt{x^3}}$

Exercise 2

(a) Rewritten as $x^{\frac{5}{2}}$. Derivative is $\dfrac{5}{2}x^{\frac{3}{2}}$ or $\dfrac{5}{2}\sqrt{x^3}$

(b) Rewritten as $x^{\frac{4}{3}}$. Derivative is $\dfrac{4}{3}x^{\frac{1}{3}}$ or $\dfrac{4}{3}\sqrt[3]{x}$

(c) Rewritten as $x^{\frac{5}{2}} + x^{\frac{1}{2}}$. Derivative is $\frac{5}{2}x^{\frac{3}{2}} + \frac{1}{2}x^{-\frac{1}{2}}$ or $\frac{5}{2}\sqrt{x^3} + \frac{1}{2\sqrt{x}}$

(d) Rewritten as $x^{-2} + x^{-\frac{9}{2}}$. Derivative is $-2x^{-3} - \frac{9}{2}x^{-\frac{11}{2}}$ or $-\frac{2}{x^3} - \frac{9}{2\sqrt{x^{11}}}$

Exercise 3

(1) $12y - x - 16 = 0$

(2) $9y + x - 8 = 0$

(3) $(16, 3)$

(4) $(5, 10)$ minimum and $(-5, -10)$ maximum

(5) $\dfrac{dy}{dx} = 3x^2 + \dfrac{1}{x^2} + \dfrac{6}{x^4}$ and each term is always positive

Integration

Key learning objectives

- to integrate terms of the form ax^n where n is any rational number except -1
- to use a range of algebraic techniques to rewrite expressions so that they can be integrated
- to solve simple differential equations to find an expression for y in terms of x
- to evaluate definite integrals — applying them to areas where required
- to learn and apply the trapezium rule to approximate an area beneath a curve; justify whether such an estimate is an under-estimate or over-estimate and identify how it can be improved

In Core 1 you learnt that when $y = ax^n$ then $\dfrac{dy}{dx} = \dfrac{ax^{n+1}}{n+1}$. In words the rule was 'add 1 to the power and then divide by the new power'. This rule was only used when n was a positive integer. By using knowledge of indices you can now extend this idea to differentiate any term where n is any rational number, excluding the value -1. As with differentiation the two crucial results are:

$$\frac{1}{x^n} = x^{-n} \text{ and } \sqrt[n]{x} = x^{\frac{1}{n}}$$

Consider the results shown below for negative powers of x:

y	y rewritten using negative powers	$\int y\,dx$	$\int y\,dx$ rewritten without negative powers
$\dfrac{1}{x^2}$	x^{-2}	$\dfrac{x^{-1}}{-1}$	$-\dfrac{1}{x}$
$\dfrac{1}{x^3}$	x^{-3}	$\dfrac{x^{-2}}{-2}$	$-\dfrac{1}{2x^2}$
$\dfrac{1}{x^n}$	x^{-n}	$\dfrac{x^{-n+1}}{-n+1}$	$-\dfrac{1}{(n-1)x^{n-1}}$

Similarly the next set of results relate to fractional powers of x:

y	y rewritten using fractional powers	$\dfrac{dy}{dx}$	$\dfrac{dy}{dx}$ rewritten without negative or fractional powers
$\sqrt{x}$	$x^{\frac{1}{2}}$	$\dfrac{2x^{\frac{3}{2}}}{3}$	$\dfrac{2}{3}\sqrt{x^3}$
$\sqrt[3]{x}$	$x^{\frac{1}{3}}$	$\dfrac{3x^{\frac{4}{3}}}{4}$	$\dfrac{3}{4}\sqrt[3]{x^4}$
$\sqrt[n]{x}$	$x^{\frac{1}{n}}$	$\dfrac{nx^{\frac{n+1}{n}}}{n+1}$	$\dfrac{n}{n+1}\sqrt[n]{x^{n+1}}$
$\sqrt[n]{x^m}$	$x^{\frac{m}{n}}$	$\dfrac{nx^{\frac{n+m}{n}}}{n+m}$	$\dfrac{n}{n+m}\sqrt[n]{x^{n+m}}$

The following examples have a mixture of terms that need to be rewritten along with some that do not.

As with Core 1, integrals can be indefinite (no limits, remember to add '+ c') or definite (limits).

Worked example 1

Obtain $\int\left(10x^4 - \dfrac{2}{x^3} + 1\right)dx$.

Solution

First rewrite the expression as $10x^4 - 2x^{-3} + 1$

Then integrate to get $\dfrac{10x^5}{5} - \dfrac{2x^{-2}}{(-2)} + x + c$, which simplifies to:

$2x^5 + x^{-2} + x + c$ or $2x^5 + \dfrac{1}{x^2} + x + c$

Worked example 2

Evaluate $\displaystyle\int_1^4 \left(\sqrt{x} - 1\right)dx$.

Solution

First rewrite to get $\displaystyle\int_1^4 \left(x^{\frac{1}{2}} - 1\right)dx$

Integrate to get $\left[\dfrac{2}{3}x^{\frac{3}{2}} - x\right]_1^4$

Substitute limits to get $\left[\dfrac{2}{3}(4)^{\frac{3}{2}} - 4\right] - \left[\dfrac{2}{3}(1)^{\frac{3}{2}} - 1\right] = \left[\dfrac{16}{3} - 4\right] - \left[\dfrac{2}{3} - 1\right] = \dfrac{4}{3} - -\dfrac{1}{3} = \dfrac{5}{3}$

Exercise 1

(1) Obtain $\int\left(8x^3 - 5x^2 + \dfrac{4}{x^2} - \dfrac{1}{x^3} \right)dx$.

(2) Obtain $\int\left(\sqrt{x} + \dfrac{4}{\sqrt{x}} \right)dx$.

(3) $\int_{1}^{4}\left(10 - \dfrac{15}{x^4} \right)dx$.

(4) $\int_{-1}^{8}\left(\sqrt[3]{x} + 4 \right)dx$.

In addition some questions may require the use of two further rules of indices:

$$x^m x^n = x^{m+n} \text{ and } \dfrac{x^m}{x^n} = x^{m-n}$$

Worked example 1

Express $x(\sqrt{x} - 3)$ in the form $x^p - x^q$ where p and q are rational numbers.

Hence obtain $\int x(\sqrt{x} - 3)dx$.

Solution

Rewrite with powers to get $x(x^{\frac{1}{2}} - 3)$

Then expand the brackets to get $xx^{\frac{1}{2}} - 3x$

Using the appropriate law of indices this becomes $x^{\frac{3}{2}} - 3x$ which is in the correct form.

Now integrate to get $\dfrac{2}{5}x^{\frac{5}{2}} - \dfrac{3}{2}x^2 + c$ or $\dfrac{2}{5}\sqrt{x^5} - \dfrac{3}{2}x^2 + c$

Worked example 2

Express $\dfrac{4x^3 + x}{x^5}$ in the form $x^p + x^q$, where p and q are integers and hence integrate the expression.

Solution

Rewrite as two fractions to get $4\dfrac{x^3}{x^5} + \dfrac{x}{x^5}$.

Use the appropriate law of indices to write as single powers of x to get $4x^{-2} + x^{-4}$ which is the required format.

Now integrate to get $\dfrac{4}{-1}x^{-1} - \dfrac{x^{-3}}{-3} + c$

Tidying up gives $-4x^{-1} + \dfrac{x^{-3}}{3} + c$ or $-\dfrac{4}{x} + \dfrac{1}{3x^3} + c$

Exercise 2

In each case rewrite in the form x^p where p is a rational number, then integrate the expression.

(a) $x^4\sqrt{x}$

(b) $(\sqrt{x})(\sqrt[3]{x})$

In each case rewrite in the form $x^p - x^q$ where p and q are rational numbers and hence integrate the expression.

(c) $\sqrt[3]{x}(\sqrt[3]{x^2} - 1)$

(d) $\dfrac{6x^3 - 1}{x^5}$

As in Core 1, questions about integration can relate to finding the equation of a curve given the gradient function (an expression for $\dfrac{dy}{dx}$) or finding an area beneath a curve or between two curves. A brief reminder about each one is given below before a number of examples are given and brief exercises follow.

Finding the equation of a curve

Integrate the given expression, remembering to include '+ c'. This finds the general solution of the curve. To find a particular solution, a point on the curve must be given in the question (when $x = ..., y = ...$). Substitute these values to find the value of the constant. The particular solution has then been found.

To find an area beneath a curve

Definite integrals are used to calculate an area bounded by a curve and the x-axis. The graph of $y = f(x)$ is shown below:

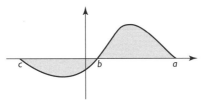

The shaded area above the x-axis is given by $\int_a^b f(x)\,dx$.

The shaded area below the x-axis is given by $\int_c^b f(x)\,dx$.

The second integral will give a negative answer, indicating it is below the x-axis.

Integration can also be used to calculate an area between two curves or a curve and a line.

As a reminder these methods are given below.

Method 1

- Find the points of intersection of the two curves — the x values will be the limits of integration.
- Use integration to calculate the area beneath each curve.
- Find the difference between these answers to obtain the required area.

Method 2

- Find the points of intersection of the two curves — the x values will be the limits of integration.
- Obtain a single algebraic expression by subtracting one curve from the other.
- Use integration with the single expression to find the required answer.

Worked example 1

Given that $\dfrac{dy}{dx} = \dfrac{10}{x^3} + 1$ and that the curve goes through the point $(1, 2)$, find y in terms of x.

Solution

First rewrite with negative powers to get $10x^{-3} + 1$

Now integrate to get $y = -5x^{-2} + x + c$ or $y = -\dfrac{5}{x^2} + x + c$

Now substitute the given point to get $2 = -\dfrac{5}{(1)^2} + (1) + c$ and hence $c = 6$

So $y = -\dfrac{5}{x^2} + x + 6$

Worked example 2

Find the finite area bounded by the curve $y = \sqrt{x} + 1$, the x-axis and the lines $x = 0$ and $x = 1$.

Solution

This area $= \displaystyle\int_0^1 (\sqrt{x} + 1)\,dx$ so rewrite with powers to get $\displaystyle\int_0^1 (x^{\frac{1}{2}} + 1)\,dx$

Integrating gives $\left[\dfrac{2}{3}x^{\frac{3}{2}} + x \right]_0^1 = \left(\dfrac{2}{3} + 1 \right) - (0) = \dfrac{5}{3}$

Worked example 3

Find the area bounded by the curves and $y = \sqrt{x}$ and $y = \sqrt[3]{x}$.

Solution

These curves intersect when $\sqrt{x} = \sqrt[3]{x}$, so raising each side to the power of 6 gives $x^3 = x^2$. Hence $x^3 - x^2 = 0$ therefore $x^2(x - 1) = 0$ so $x = 0$ and $x = 1$ are the intersection points.

Required area $= \displaystyle\int_0^1 (\sqrt[3]{x} - \sqrt{x})\,dx$

Rewrite with fractional powers:

$\displaystyle\int_0^1 (x^{\frac{1}{3}} - x^{\frac{1}{2}})\,dx$ then integrate to get $\left[\dfrac{3}{4}x^{\frac{4}{3}} - \dfrac{2}{3}x^{\frac{3}{2}} \right]_0^1 = \left(\dfrac{3}{4} - \dfrac{2}{3} \right) - (0) = \dfrac{1}{12}$

Exercise 3

(1) Given that $\dfrac{dy}{dx} = \dfrac{1}{\sqrt{x}}$ and that the curve goes through the point $(9, 10)$, find y in terms of x.

(2) Find the finite area bounded by the curve $y = 1 + \dfrac{9}{x^2}$, the x-axis and the lines $x = 1$ and $x = 2$.

(3) Determine the points of intersection of the curves $y = 3\sqrt{x}$ and $y = 2x$. Hence find the area bounded by the curves $y = 3\sqrt{x}$ and $y = 2x$.

The trapezium rule

Some areas cannot yet be found using integration, because the function for the curve involved cannot be integrated using knowledge from Core 1 or Core 2. In such cases use the trapezium rule to obtain an approximate answer. If you are required to do this, the question will specifically request it.

The basic idea is that you divide the area required into several trapezia, each of the same width, and then add the areas of these together. It is possible to make a general formula — stated in the formula book. For notation purposes we use $x_0, x_1, x_2, x_3, \ldots$ to denote the x values at which each trapezium starts and ends. Then $y_0, y_1, y_2, y_3, \ldots$ denote the corresponding heights of the parallel sides in each trapezium. Use h for the width of each trapezium (same for each trapezium).

A brief explanation of this formula is given below. In this case there are just two trapezia.

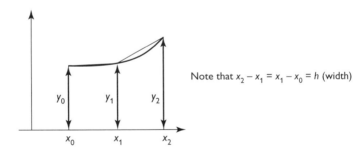

Note that $x_2 - x_1 = x_1 - x_0 = h$ (width)

The formula for the area of a trapezium is $\dfrac{1}{2}$ (sum of heights of parallel sides)(width between them)

So the area of the first trapezium $= \dfrac{1}{2}(y_0 + y_1)(h)$

And the area of the second trapezium $= \dfrac{1}{2}(y_1 + y_2)(h)$

Adding these together and reordering gives area $= \dfrac{1}{2}(y_0 + 2y_1 + y_2)(h)$

Notice than the middle height has been counted twice because it occurs in both trapezia. The 'outside' heights have been counted only once each. By considering this structure the formula for n strips can be generalised as:

$$\text{area} = \frac{1}{2}h[y_0 + y_n + 2(y_1 + y_2 + y_3 + \ldots + y_{n-1})]$$

where $h = \dfrac{b-a}{n}$ (b = upper x limit, a = lower x limit)

This formula can give an over-estimate or under-estimate depending on the shape of the curve. For example:

This shape results in an **over-estimate** since the top of each trapezium is **above** the curve.

This shape results in an **under-estimate** since the top of each trapezium is **below** the curve.

Worked example

Use the trapezium rule with four strips to estimate $\int_{1}^{2} 2^x dx$ correct to 3 significant figures.

Solution

First calculate $h = \dfrac{2-1}{4} = 0.25$, therefore the x values will be 1, 1.25, 1.5, 1.75 and 2.

Use these to work out the y values as shown in the list below:

$x_0 = 1$ $y_0 = 2^1 = 2$
$x_1 = 1.25$ $y_1 = 2^{1.25} = 2.3784\ldots$
$x_2 = 1.5$ $y_2 = 2^{1.5} = 2.8284\ldots$
$x_3 = 1.75$ $y_3 = 2^{1.75} = 3.3635\ldots$
$x_4 = 2$ $y_4 = 2^2 = 4$

Notice that any decimal values must have more figures than the final required accuracy.
Now substitute in the formula to get:

Estimate $= \dfrac{1}{2} \times 0.25[2 + 4 + 2(2.3784 + 2.8284 + 3.3635)] = 2.89$

Expect there to be 2, 3 or 4 strips which imply 3, 4 or 5 x and y values respectively (for n strips there will be $n + 1$ such values). The x values are often referred to as ordinates.

Exercise 4

In each case use the trapezium rule to estimate the value of the integral correct to 3 significant figures.

(a) $\int_{1}^{3} (\sqrt{x} + 1)dx$

(b) $\int\limits_0^1 3^x \, dx$

(c) $\int\limits_0^2 \dfrac{1}{x^3 + 1} \, dx$

Answers to exercises: integration

Exercise 1

(1) $2x^4 - \dfrac{5}{3}x^3 - 4x^{-1} + \dfrac{1}{2}x^{-2} + c$ or $2x^4 - \dfrac{5}{3}x^3 - \dfrac{4}{x} + \dfrac{1}{2x^2} + c$

(2) $\dfrac{2}{3}x^{\frac{3}{2}} + 8x^{\frac{1}{2}}$ or $\dfrac{2}{3}\sqrt{x^3} + 8\sqrt{x}\ \ + c$

(3) $\dfrac{1605}{64}$

(4) $\dfrac{189}{4}$

Exercise 2

(a) Rewrite as $x^{\frac{9}{2}}$ and integrate to get $\dfrac{2}{11}x^{\frac{11}{2}} + c$ or $\dfrac{2}{11}\sqrt{x^{11}} + c$.

(b) Rewrite as $x^{\frac{5}{6}}$ and integrate to get $\dfrac{6}{11}x^{\frac{11}{6}} + c$ or $\dfrac{6}{11}\sqrt[6]{x^{11}} + c$.

(c) Rewrite as $x - x^{\frac{1}{3}}$ and integrate to get $\dfrac{x^2}{2} - \dfrac{3}{4}x^{\frac{4}{3}} + c$ or $\dfrac{x^2}{2} - \dfrac{3}{4}\sqrt[3]{x^4} + c$.

(d) Rewrite as $6x^{-2} - x^{-5}$ and integrate to get $-6x^{-1} + \dfrac{x^{-4}}{4} + c$ or $-\dfrac{6}{x} + \dfrac{1}{4x^4} + c$.

Exercise 3

(1) $y = 2x^{\frac{1}{2}} + 4$ or $y = 2\sqrt{x} + 4$

(2) $\dfrac{11}{2}$

(3) Points are $(0, 0)$ and $(\frac{9}{4}, \frac{9}{2})$, area $= \dfrac{27}{16}$

Exercise 4

(a) 3.45

(b) 1.83

(c) 1.09

Questions
&
Answers

In this section of the guide there are nine questions based on the topic areas outlined in the Content Guidance section. The section is structured as follows:

- **sample questions** in the style of the module
- **example candidate responses at the C/D boundary (Candidate A)**, which demonstrate some sound application of knowledge but contain weaknesses, including careless errors. There is much potential for improvement.
- **example candidate responses at the A/B boundary (Candidate B)**, which demonstrate thorough knowledge and good application of key results. Some answers show a novel insight and the solution is obtained efficiently, but some crucial marks have been lost.

Examiner's comments

All candidate responses are followed by examiner's comments, denoted by the icon [e]. These indicate where credit has been awarded and where marks have been lost. They point out areas for improvement, specific problems, and common errors made by candidates.

Examiners award different types of mark according to the agreed mark scheme. These are referred to in the commentary. They are:

- **method marks (M)** — awarded for an attempt to apply a correct method
- **accuracy marks (A)** — awarded for a correct answer, but only if the method mark has been obtained
- **explanation marks (E)** — awarded for a correct explanation
- **independent marks** — a combination of method and accuracy, often for stand-alone answers
- **follow-through marks** — credit is given for using an earlier incorrect answer correctly in a later part of a question. On exam papers, such marks are limited.

Question 1

Limit of a sequence

The nth term of a sequence is u_n. The sequence is defined by $u_{n+1} = au_n + b$ where a and b are constants.

Given that $u_1 = 40, u_2 = 30, u_3 = 35$:

(a) Show that $a = -\dfrac{1}{2}$ and find the value of b. (5 marks)

(b) Find the value of u_4. (1 mark)

(c) The limit of u_n as n tends to infinity is L. Write down an equation for L and hence, or otherwise, find the exact value of L. (3 marks)

Total: 9 marks

Candidates' answers to Question 1

Candidate A

(a) When $a = -\frac{1}{2}$ then $40\left(-\frac{1}{2}\right) + b = 30$

so $-20 + b = 30$ and $b = 50$

⟳ The candidate tries to use the given answer but uses only u_1 and u_2 to do this. However, the correct value of b is obtained. This kind of solution is often referred to as a 'special case' and can score only a limited number of marks. On this occasion 2 marks are awarded (M1A1).

(b) $35\left(-\frac{1}{2}\right) + 50 = 32.5$

⟳ The value is correct and the recurrence relation is used with the value of b found in part (a), earning the mark.

(c) 40 30 35 32.5...33.75 33.125 33.4375 this gives a limit of 33.333... = $33\frac{1}{3}$

⟳ The candidate does not form the requested equation and has clearly used a calculator correctly to generate the correct sequence and has recognised the exact answer. Only 1 mark is lost for not stating the required equation. A further mark would have been lost if an exact answer had not been given.

⟳ **The candidate's total mark is 5 out of 9, which is a borderline C/D-grade answer. Some understanding of recurrence relations is shown but it is not thorough enough to score more marks.**

Candidate B

(a) $30 = 40a + b$

$35 = 30a + b$

Subtracting gives $-5 = 10a$ so $a = -\frac{1}{2}$

Therefore $30 = 40 \left(-\frac{1}{2}\right) + b$

$30 = -20 + b$

$b = 10$

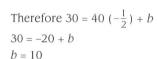 The method is clear, the candidate uses the recurrence relation with all three given values. He/she then tries to solve the simultaneous equations. The only error is at the end, where b should be equal to 50. The final accuracy mark is lost, so **4 out of 5** marks are awarded.

(b) $35\left(-\frac{1}{2}\right) + 10 = -7.5$

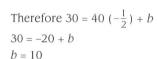 The candidate uses the recurrence relation correctly but with the incorrect value of b. The mark is awarded as a 'follow-through' mark. With the correct value of b, $u_4 = 32.5$.

(c) The limit occurs when two adjacent terms give the same answer so:

$L = L\left(-\frac{1}{2}\right) + 10$

$1.5L = 10$

$L = \dfrac{20}{3}$

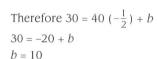 The structure of the equation is correct, but the wrong value of b has again been used. Therefore, the final value of L is incorrect, too. The candidate can score M1A1 (as a 'follow through'), but the final mark is for a fully correct answer.

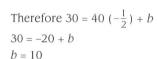 **The candidate's total mark is 7 out of 9, which is borderline grade A/B grade. A GCSE error cost him/her 2 marks.**

Question 2

Binomial expansion

(a) Given that $(1 - 3x)^4 = 1 - 12x + ax^2 + bx^3 + 81x^4$, find the values of the integers a and b. (3 marks)

(b) Find the coefficient of x^2 in the expansion of $(1 + x)^7$. (2 marks)

(c) Find the coefficient of x^2 in the expansion of $(1 + x)^7(1 - 3x)^4$. (4 marks)

Total: 9 marks

Candidates' answers to Question 2

Candidate A

(a) $(1 - 3x)^4 = 1 + 4.-3x + 6.-3x^2 + 4.-3x^3 + 1.-3x^4 = 1 - 12x - 18x^2 - 12x^3 + 81x^4$

ℓ Weaker candidates often leave out brackets in their working and this candidate makes two errors as a result of this. The structure of the binomial expansions appears to be understood from the first line, but '6.–3x^2' has been evaluated as $-18x^2$, instead of $54x^2$, and '4.–3x^3' should be $-108x^3$. The candidate scores 1 method mark and no accuracy marks.

(b) $(1 + x)^7 = 1 + 7x + \dfrac{7.6}{2}x^2 + \dots$ so $21x^2$

ℓ The candidate expands the bracket up to the required term. He/she states the *term*, not just the *coefficient* of 21. Sometimes this can be judged as 'benefit of doubt' and no marks are lost. On this occasion both marks are given, but it is not always guaranteed.

(c) Coefficient of x^2 from part a) is $-18x^2$. Times by 1 is still -18
Coefficient of x^2 from part b) is $21x^2$. Times by 1 is still 21
so total is $21 - 18 = 3x^2$

ℓ The candidate shows some understanding here, by using the x^2 terms with the constant from the other bracket. However, he/she has not considered that x^2 could be obtained from using the x terms in each bracket. He/she also uses an incorrect answer from part (a). However, the candidate can still score M1A1 but has lost 2 marks.

ℓ **The candidate's total mark is 5 out of 9 — a grade-C/D answer. His/her understanding is not fully sound and crucial numerical errors are made.**

Candidate B

(a) $(1 - 3x)^4 = 1 + 4(-3x) + 6(-3x)^2 + 4(-3x)^3 + 1(-3x)^4$
$= 1 - 12x + 54x^2 - 108x^3 + 81x^4$
so $a = 54$, $b = -108$

ℓ This is an excellent answer — brackets have been used to ensure that there are no errors with negatives. The candidate's use of coefficients from Pascal's triangle is clearly efficient. All 3 marks are scored.

(b) $\binom{7}{2}(1)^5(x)^2 = 21x^2$

coefficient $= 21$

This is another excellent response — the candidate uses the binomial coefficient method efficiently. Both marks are given.

(c) $(1 - 12x + 54x^2 \ldots)(1 + 7x + 21x^2 \ldots)$
so to get x^2 we use $-12(7) + 54(1) + 1(21)$ to get -11

The method is well understood and the candidate correctly identifies all the possible correct pairings. However, the total should be -9. Only 1 mark is lost.

The candidate's total mark is 8 out of 9. This is a high-quality, grade-A answer with only one numerical error.

Question 3

Sectors — arc length and area

The diagram shows a sector of a circle whose centre is **O** and radius is **18 cm**. A line is drawn from **A** to **B** to form a triangle **OAB**.

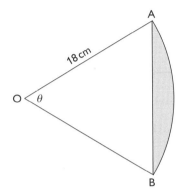

The length of the arc **AB** is **24 cm**. Angle **AOB** is θ radians.

(a) Find θ.   (2 marks)

(b) (i)   Find the area of the sector.   (2 marks)

     (ii) Hence obtain the area of the shaded segment, giving your answer to the
           nearest cm². (3 marks)

(c) Show that the side **AB = 22.3 cm**.   (2 marks)

Total: 9 marks

Candidates' answers to Question 3

Candidate A

(a) Using $s = r\theta$

$\theta = 24/18 = 1.33°$

✐ Both marks are awarded, as the method is correct and the answer is given to 3 significant figures (standard accuracy). The candidate thinks incorrectly that the answer is in degrees, but no marks are lost.

(b) (i)   Area $\frac{1}{2}r^2\theta$

so $\frac{1}{2} \times 18 \times 1.33^2 = 287$ cm²

✐ The correct formula has been stated but it is applied incorrectly. No marks are awarded because the second line does not resemble the formula that should have been used. No marks are given for simply stating a formula.

(ii) Shaded area = sector area – triangle area

triangle = $\frac{1}{2}$ (18)^{2}sin 1.33 = 3.76

shaded area = 287 – 3.76 = 283.24 cm^2

> The candidate uses the correct formula for the triangle, but treats the 1.33 as degrees not radians on the calculator. He/she also uses the incorrect answer from part (i). This latter error can be accepted as a 'follow-through' mark, so the candidate earns 2 out of 3 marks.

(c) 1.33/2 = 0.665
18sin 0.665 = 0.2089
so AB = 2 × 0.2089 = 0.4178 not as stated.

> The candidate applies knowledge of trigonometry in a right-angled triangle. The only error is again the use of degrees and not radians. 1 method mark is awarded.

> **The candidate's total mark is 5 out of 9 — a grade-C/D answer. Correct use of radians would have increased the total by 2 marks, making the answer worthy of a B grade.**

Candidate B

(a) Using $s = r\theta$

$\theta = 24 \div 18 = \frac{4}{3}$

> Both marks are awarded, as the method is correct and the answer is given as an exact fraction.

(b) (i) Area = $\frac{1}{2} r^2 \theta$

so $\frac{1}{2} \times 18^2 \times \frac{4}{3}$ = 216 cm^2

> This is an excellent response. The answer is exact and the working is clearly shown. Both marks are awarded.

(ii) Triangle = $\frac{1}{2}$ (18)^2 sin $\frac{4}{3}$ π = 140.3

Shaded area = 216 – 140.3 = 75.7 cm^2

> The correct triangle formula has been used, but the angle of $\frac{4}{3}$ has become $\frac{4}{3}$ π. Quite often candidates inexplicably add a π for no valid reason. 1 mark is lost for this error. The correct answer is 216 – 157.5 = 58.5.

(c) Use the cosine rule

AB^2 = 18^2 + 18^2 – 2(18)(18)cos $\frac{4}{3}$ π = 972

So answer = 31.2 cm

The same error has been made as in the previous part, so 1 mark is still lost, since the final answer does not match the printed answer.

The candidate's total mark is 7 out of 9. This is a borderline grade-A/B response. It is only careless errors that have lost marks — understanding is good.

Question 4

Manipulation of logarithms

(a) Given that the constant k satisfies:

$2\log_a k - \log_a (3k + 20) = \log_a 2$

    (i) Show that $k^2 - 6k - 40 = 0$.   **(3 marks)**

    (ii) Deduce that there is only one possible value for k and find its value.   **(3 marks)**

(b) **(i)** Given that $\log_a p = 3\log_a 4 - \dfrac{1}{2}\log_a 16$, show that $p = 16$.   **(3 marks)**

    (ii) Given that $\log_a q = 10 - \dfrac{1}{2}\log_a 25$, express q in terms of a.

    There must not be any logs in your final answer.   **(3 marks)**

Total: 12 marks

Candidates' answers to Question 4

Candidate A

(a) **(ii)** $\quad k^2 - 6k - 40 = 0$

$\qquad\qquad (k + 4)(k - 10) = 0$

$\qquad\qquad$ so $k = -4$ or 10 so two values

 This candidate chooses to answer part (ii) first and identifies two solutions. However, he/she fails to see that one of them is not valid (only $k = 10$ is valid because if $k = -4$, $\log_a (-4)$ is not valid). One mark is lost.

(a) **(i)**   Substituting $k = 10$ we get $2\log_a 10 - \log_a(50) = 0.301\ldots = \log_a 2$. So one value works.

 This response scores no marks. The candidate tries to recover part (i) by using part (ii). He/she has used numerical values on the calculator, in base 10, so there is no real manipulation of logs. The other k value has not been considered.

(b) **(i)** $3\log_a 4 = \log 64$

$\qquad\qquad \dfrac{1}{2}\log_a 16 = \log 4$

$\qquad\qquad \log 64 - \log 4 = \log 16$ so $p = 16$

 There appears to be some log manipulation here, although the base appears to have been lost. The candidate does not explain exactly how the manipulation has been done. To some extent 'benefit of the doubt' would apply, although at most 2 marks would be awarded as the solution is far from perfect.

(b) **(ii)** $\log q + \dfrac{1}{2}\log_a 25 = \log 5q$ so $a = 5q$

e This is incorrect but the manipulation to get to log 5q is just worthy of 1 mark, as it moves towards the correct solution. It is possible that base 10 has been used again, but progress has been made nevertheless.

e **The candidate's total mark is 5 out of 12 — a grade-E response. Manipulation of logs is a difficult area for weaker candidates. They tend to score the odd mark or try to use decimal values — to avoid such errors, learn all the laws of logs.**

Candidate B

(a) (i) Using laws of logs, $2\log_a k = \log_a k^2$ and $\log_a 2 + \log_a (3k + 20) = \log_a (6k + 40)$
Therefore $\log_a k^2 = \log_a (6k + 40)$ means $k^2 = 6k + 40$
Hence $k^2 - 6k - 40 = 0$

e This is an excellent solution — each step is justified. Division is avoided by moving one of the log terms across to the right-hand side first. Full marks are awarded.

(ii) Factorise $k^2 - 6k - 40 = 0$ to get $(k + 4)(k - 10) = 0$ so $k = -4$ or $k = 10$ — both work

e The candidate scores 2 marks. He/she has not ruled out $k = -4$. He/she should have realised that $\log_a k$ is undefined if $k = -4$.

(b) (i) $\log_a p + \frac{1}{2} \log_a 16 = \log_a p + \log_a 4 = \log_a 4p$

$3\log_a 4 = \log_a 4^3 = \log_a 64$
Therefore $4p = 64$ so $p = 16$

e This is another excellent response. The candidate rearranges and works with each separate expression to give a well-explained solution, which earns full marks.

(ii) $\log_a q + \frac{1}{2} \log_a 25 = \log_a q + \log_a 5 = \log_a 5q$

If $\log_a 5q = 1$ then $5q = a$ so $q = \frac{a}{5}$

e The candidate builds on the ideas from (b)(i) and scores full marks for this part.

e **The candidate's total mark is 11 out of 12 — easily a grade-A response. Manipulation of logs is clearly a strong area for this candidate.**

Question 5

Geometric series

The third term of a geometric series is 4 and the fifth term is $\frac{1}{4}$.

(a) Show that $r^2 = \frac{1}{16}$. (3 marks)

(b) Deduce that there are two possible sequences, stating the first term and common ratio of each one. (3 marks)

(c) Find the sum to infinity of each of the series in part (b). (3 marks)

Total: 9 marks

Candidates' answers to Question 5

Candidate A

(a) $4 \times \frac{1}{16} = \frac{1}{4}$ and there are two common ratios from third to fifth term so $r^2 = \frac{1}{16}$ works.

e This answer is not well explained, but it shows a limited understanding of why the statement is true. On balance, it is worth 1 mark as a special case. Had the candidate argued that $4r^2 = \frac{1}{4}$ and then ended with the statement given, full marks could have been awarded.

(b) $r = \frac{1}{4}$ or $-\frac{1}{4}$ hence two sequences

So $4 \div \frac{1}{4} = 16$ and -16 are the first terms.

e The r values are correct but the candidate has only divided by the ratio and not the ratio squared, so the a values are incorrect. He/she scores 2 marks.

(c) $a = 16$ and $r = \frac{1}{4}$ gives $S_\infty = \dfrac{16}{1 - \frac{1}{4}} = \dfrac{64}{3}$

Other answer $= -\dfrac{64}{3}$

e The candidate uses the incorrect answers from part (b) correctly to get the first sum to infinity. The second is incorrect. He/she earns 2 marks out of 3 (M1A1 as a follow through).

e **The candidate's total is 5 marks out of 10. This is a borderline grade-D/E answer. Series questions often cause difficulties if practice has been limited to typical textbook questions and simple use of formulae.**

Candidate B

(a) $ar^2 = 4$

$ar^4 = \frac{1}{4}$

Comparing then $r^2 = \frac{1}{16}$

e The candidate makes good use of the nth term formulae for geometric series, but 'comparing' does not fully explain how to get the result. Hence, 2 marks out of 3 are awarded (M1A1A0).

(b) Take square root so $r = \frac{1}{4}$ or $-\frac{1}{4}$

From part (a) $a = \frac{4}{r^2}$

So $a = 64$ in both cases

e The candidate makes good use of formulae — rearranging before substituting. This saves additional work. All 3 marks are scored.

(c) $S_\infty = \frac{a}{1-r}$

so $S_\infty = \frac{64}{1-\frac{1}{4}} = \frac{256}{3}$

or $S_\infty = \frac{64}{1+\frac{1}{4}} = \frac{256}{5}$

e This is an excellent response. The candidate states the formula clearly before using it and earns full marks.

e **The candidate's total is 8 marks out of 9. This is a grade-A answer. Lack of clarity in one part lost a single mark.**

Question 6

Arithmetic series

The first term of an arithmetic series is a and the common difference is d. The fifth term is 0 and the thirtieth term is -75.

(a) (i) Show that $a + 4d = 0$ and state one equation connecting a and d. (2 marks)

    (ii) Hence, or otherwise, show that $a = 12$ and find the value of d. (2 marks)

(b) Obtain $\sum_{n=1}^{20} 15 - 3n$. (2 marks)

(c) Find the least value of k such that $\sum_{n=1}^{k} 15 - 3n < -1500$. (4 marks)

Total: 10 marks

Candidates' answers to Question 6

Candidate A

(a) (i) nth term is $a + (n - 1)d$

So $0 = a + (5 - 1)d$

$a + 4d = 0$

$a + (30 - 1)d = -75$

$a + 29d = -75$

e Candidate A scores both marks. The correct formula has been stated and used. He/she shows clearly how to derive the equations.

(ii) $25d = -75$ so $d = -3$ and $a = -12$

e The correct value for d is found, but the candidate does not show how to obtain the value of a. A single mark is awarded.

(b) $a = 12, d = -3, n = 20$

$S_n = \frac{n}{2}[2a + (n - 1)d]$

$S_n = \frac{20}{2}[2(12) + (20 - 1)(-3)]$

$= 183$

e The candidate states the correct formula and substitutes the right values, but makes an error with the evaluation. The correct answer is -330. A single method mark is awarded.

(c) $S_{20} = 183$

$d = -3$

$-1500 - 183 = -1683$

$-1683 \div -3 = 561$ terms

e This answer does not score any marks. The method does not work, even if S_{20} were correct.

e **The candidate's total mark is 4 out of 10 — a grade-E answer. A few care-less slips lost 2 crucial marks in particular, which cost the candidate a grade C. Part (c) is a discriminating question for the higher grades.**

Candidate B

(a) (i) $n = 5$ so $a + 4d = 0$

$n = 30$ so $a + 29d = -75$

e The candidate does not explain this well, since one answer is given in the question. One mark is awarded. He/she should have stated the general formula first.

(ii) Subtract the equations so $25d = -75$ and $d = -3$, hence $a = 12$

e Both values are correct and the candidate earns 2 marks.

(b) $S_{20} = -330$

e No working is shown but both marks are awarded for a correct answer. The candidate may have added terms together or used the formula on a calculator — a risky strategy.

(c) $S_k = \dfrac{k}{2}[24 - 3(k - 1)] < -1500$

$k(27 - 3k) < -3000$

$0 < 3k^2 - 27k - 3000$

$0 < k^2 - 9k - 1000$

$k = \dfrac{9 \pm \sqrt{9^2 - 4(1)(-1000)}}{2}$

$k = 36.44$

e This is a good solution as an inequality is formed initially. The final mark is lost because an integer is needed (term positions must be integers). To be less than -1500, $n = 37$ (round up to the nearest integer).

e **The candidate's total mark is 8 out of 10 — just on the grade-A/B border. Only minor slips are in evidence. The answer to the last part is indicative of a grade-A candidate.**

Question 7

Exponential graphs and numerical methods

The diagram shows a sketch of the curve with equation $y = 4(3^x) - 1$.

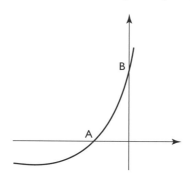

The curve intersects the x-axis at **A** and the y-axis at **B**.

(a) (i) State the y-coordinate of **B**. (1 mark)

    (ii) Find the x-coordinate of **A** correct to 3 significant figures. (3 marks)

(b) (i) Use the trapezium rule with four strips to find an approximate value

$$\text{for } \int_{0}^{2} \left[4(3^x) - 1 \right] dx$$

Give your answer correct to 3 significant figures. (4 marks)

    (ii) With the aid of a sketch, explain whether your answer is an under-estimate

or an over-estimate for the exact value of $\int_{0}^{2} \left[4(3^x) - 1 \right] dx$. (3 marks)

(c) The curve $y = 4(3^x)$ can be transformed into the curve $y = 4(3^x) - 1$ with
the aid of a single transformation. State this transformation. (1 mark)

Total: 12 marks

Candidates' answers to Question 7

Candidate A

(a) (i) When $x = 0$, $y = 4 \times 3 - 1 = 11$

 This answer is incorrect and does not earn the mark, as the candidate appears to think
that $3^0 = 3$.

    (ii) $4(3^x) - 1 = 0$. Using my graphical calculator $x = -1.26$

 The answer is correct to 3 significant figures and all 3 marks have to be awarded.
The candidate has clearly learnt how to use the trace/zoom/solver functions on the
graphical calculator.

(b) (i) $x_0 = 0$ $y_0 = 3$

        $x_1 = 0.5$ $y_1 = 5.92$

        $x_2 = 1$ $y_2 = 11$

        $x_3 = 1.5$ $y_3 = 19.8$

        $x_4 = 2$ $y_4 = 35$

        Area $= \frac{1}{2} \times 0.5 \times [3 + 35 + 2(5.92 + 11 + 19.8)]$

             $= 20.2$

e   First, the correct y value for $x = 0$ has been obtained. The values stated for y_n are rounded to 3 significant figures, which means that the final answer is not to the required accuracy. Premature rounding is an error that is penalised. Furthermore, although the structure of the formula is correct, the evaluation is not, so another error has been made. This attempt is awarded M1A1 for correct formula and substitution.

    (ii) It will be an over-estimate because the top of each trapezium is above the curve, adding more to the area.

e   No sketch is given and the explanation lacks some clarity. However, on balance 1 mark would be awarded, since over-estimate is correct.

(c) Move down vertically by 1 unit.

e   This is correct for 1 mark, although 'translated vertically' would be a better response.

e   **The candidate's total mark is 7 out of 12 — borderline C/D-grade answer. Greater care with numerical working would have added a few crucial marks. These are typical slips by a grade-D candidate.**

Candidate B

(a) (i) $x = 0, y = 3$

e   This correct response earns the mark.

    (ii) $4(3^x) = 1$

        $3^x = 0.25$

        $\log 3^x = \log 0.25$

        $x = \dfrac{\log 0.25}{\log 3} = -1.26$

e   This well-explained response shows that the candidate understands how to work with logs. All 3 marks are awarded.

(b) (i) $\frac{1}{2} \times 0.5[\, 3 + 35 + 2\{\, 4(3^{0.5}) - 1 + 11 + 4(3^{1.5}) - 1\}]$

        $= 25.6$

e   Candidate B uses the correct formula with exact answers, which is fine. However, the evaluation is incorrect — it is a risk to enter everything at once, with so many brackets involved. The candidate loses 1 mark for this slip. The correct answer is 27.9.

(ii)

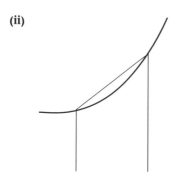

Additional area counted so it is an over-estimate.

ℓ All 3 marks are awarded — the simple diagram aids explanation.

(c) The curve is stretched by a factor 4 and then translated by 1 unit.

ℓ This response does not score the mark. The candidate has mistakenly described two transformations. He/she also misses out the direction.

ℓ **The candidate's total mark is 10 out of 12 — just into the grade-A category. He/she lost 2 marks, the second one largely for not reading the question carefully.**

Question 8

Solving trigonometrical equations

(a) **Show that $2\tan\theta\sin\theta + 3 = 0$ can be written as $2\cos^2\theta - 3\cos\theta - 2 = 0$.** (3 marks)

(b) (i) **By solving $2\cos^2\theta - 3\cos\theta - 2 = 0$, deduce that there is only one value of $\cos\theta$ for which the equation is true.** (3 marks)

    (ii) **Hence solve $2\tan\theta\sin\theta + 3 = 0$, giving your solutions in the range $0 \le \theta \le 360°$.** (2 marks)

(c) **Write down all the values of x in the interval $0 \le x \le 120°$ for which $2\tan 3x\sin 3x + 3 = 0$.** (2 marks)

Total: 10 marks

Candidates' answers to Question 8

Candidate A

(a) $\tan\theta\sin\theta = -1.5$ works if $\theta = 120°$
Then $\cos^2 120° - 3\cos 120° - 2 = 0$ from calculator.

No marks are scored because the candidate has not answered the question and has essentially spotted solutions.

(b) (i) Factorising gives $(2\cos\theta + 1)(\cos\theta - 2) = 0$

so $\cos\theta = -\frac{1}{2}$ or $\cos\theta = 2$

$120°$ works so $\cos\theta = -\frac{1}{2}$ must be the one.

The candidate factorises and solves the equation correctly but does not explain why there is only one solution. 2 out of 3 marks.

    (ii) $120°$ must work and $240°$ as well.

Both marks are awarded for two correct solutions.

(c) $40°$ by trial and error

The candidate earns 1 mark for one correct solution.

The candidate's total mark is 5 out of 10 — approximately a grade D. He/she has limited knowledge of solving trigonometrical equations and is not confident in using identities. This is typical of a grade-D/E candidate.

Candidate B

(a) $\tan\theta\sin\theta = \dfrac{\sin^2\theta}{\cos\theta}$

$2\dfrac{\sin^2\theta}{\cos\theta} + 3 = 0$

So $2\sin^2\theta + 3\cos\theta = 0$

$2(\cos^2\theta - 1) + 3\cos\theta = 0$
$2\cos^2\theta - 2 + 3\cos\theta = 0$
$2\cos^2\theta - 3\cos\theta - 2 = 0$

✎ Candidate B makes a fair attempt and has some knowledge of identities. However, he/she makes an error by thinking that $\sin^2\theta = \cos^2\theta - 1$. Nevertheless, the answer is still worth 2 out of the 3 marks (most likely M1M1A0).

(b) (i) Factorising gives $(2\cos\theta + 1)(\cos\theta - 2) = 0$
So $\cos\theta = -\frac{1}{2}$ or $\cos\theta = 2$
The second value is not possible because the maximum value of $\cos\theta$ is 1.

✎ All 3 marks are awarded for a correct solution and a sound explanation.

(ii) For $\cos\theta = -\frac{1}{2}$ then $\theta = 210°$ and $330°$.

✎ The candidate makes a slip here — possibly by using the $\sin^{-1}$ button rather than $\cos^{-1}$. No marks are awarded.

(c) Divide by 3 since $3x = \theta$ so $70°$ and $110°$.

✎ Both marks are awarded — the answers are incorrect only because of the error in (b)(ii). These marks are therefore 'follow-through' marks. The correct answers are $40°$ and $80°$.

✎ **The candidate's total mark is 7 out of 10 — a grade-B response. Pressing the wrong key on the calculator lost him/her 2 marks. Careless errors cost marks and A grades. The identity work is more typical of a grade-A candidate, despite the slip.**

Question 9

Differentiation and integration

Part of the curve $y = 4x + 1 + \dfrac{54}{x^2}$ is shown below.

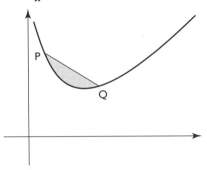

(a) Find the expression for $\dfrac{dy}{dx}$. (3 marks)

(b) (i) **Q** is a minimum point. Show that the x-coordinate of **Q** satisfies
$4x^3 = 108$. (2 marks)

   (ii) Hence deduce the coordinates of **Q**. (2 marks)

The point **P** lies on the curve and has coordinates $(1, k)$.

(c) (i) State the value of k. (1 mark)

   (ii) Find the equation of the normal to the curve at **P**, leaving your answer
   unsimplified. (4 marks)

(d) (i) Obtain $\displaystyle\int \left(4x + 1 + \dfrac{54}{x^2} \right) dx$. (3 marks)

   (ii) Find the area bounded by the arc **PQ** and the straight line **PQ** (shaded). (4 marks)

Total: 19 marks

Candidates' answers to Question 9

Candidate A

(a) $\dfrac{dy}{dx} = 4 - \dfrac{108}{x}$

Candidate A tries to differentiate but makes a common error. In differentiating
$\dfrac{54}{x^2} = 54x^{-2}$, he/she takes one less than -2 to be -1. However, the candidate still
scores 2 marks (M1A1A0), since part of the expression is correct. There is no penalty
for missing the arbitrary constant.

(b) (i) When $\dfrac{dy}{dx} = 0$

   Then $\dfrac{dy}{dx} = 4 - \dfrac{108}{x} = 0$

   So $4x = 108$ almost right!

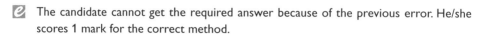

e The candidate cannot get the required answer because of the previous error. He/she scores 1 mark for the correct method.

(ii) Using $4x^3 = 108$

$x^3 = 27$

$x = 3$

$y = 19$

e Both marks are awarded. The candidate has used the printed answer to get the x value, rather than the incorrect answer obtained earlier.

(c) (i) When $x = 1$, $y = 59$, so $k = 59$

e The mark is awarded — 'state' requires no working to be shown.

(ii) Using $y - y_1 = m(x - x_1)$

$m = \text{grad} = 4 - \dfrac{108}{1} = -104$

So $y - 59 = -104(x - 1)$

$y - 59 = -104x + 104$

$y = -104x - 63$

e The candidate uses an incorrect expression for $\dfrac{dy}{dx}$, but this is due to an earlier error. The main issue is that the gradient of the tangent is used instead of the gradient of the normal. Furthermore, additional working has been done at the end that is unnecessary on this occasion (although there is no penalty for this). The candidate is awarded a total of 2 marks. The final A1 mark is lost because the rule of 'correct answer only' applies here.

(d) (i) $\displaystyle\int\left(4x + 1 + \dfrac{54}{x^2}\right)dx = 2x^2 + 1 - \dfrac{18}{x^3}$

e This attempt scores 1 mark. There are two errors: the 1 should be integrated to get x and $\dfrac{54}{x^2}$ should be integrated to get $-\dfrac{54}{x}$.

(ii) Below curve $= \left[2x^2 + 1 - \dfrac{18}{x^3}\right]_1^3$

$= \left[2(3)^2 + 1 - \dfrac{18}{(3)^3}\right] - \left[2(1)^2 + 1 - \dfrac{18}{(1)^3}\right]$

$= 18\dfrac{1}{3} - 15$

$= 3\dfrac{1}{3}$

Trapezium $= \dfrac{1}{2}(19 + 59) = 39$

So shaded area $= 39 - 3\dfrac{1}{3} = 35\dfrac{2}{3}$

e The candidate tries to apply the correct method but makes several errors. He/she substitutes into a previous wrong answer and also makes an error with the second bracket, which has a value of 15 (two negatives together). The formula for the trapezium is missing the height of 2. Candidate A scores 2 marks: M1A1 (integration and substitution).

e The candidate's total mark is **11 out of 19** — a borderline grade-C/D response. A typical grade-D response often has an early error costing marks but would not be over-penalised. A number of careless slips often distinguish grade-D responses from grade-C ones.

Candidate B

(a) $y = 4x + 1 + 54x^{-2}$

$\dfrac{dy}{dx} = 4 - 108x^{-3}$

e All 3 marks are scored. The candidate has clearly rewritten the equation before differentiating. The answer is left with a negative power, which is fully correct.

(b) (i) When $\dfrac{dy}{dx} = 0$

$4 - 108x^{-3} = 0$

$4 = 108x^{-3}$

So $4x^3 = 108$

e This response scores 1 mark only, as no explanation is given between the last two lines and the final line, which is a printed answer. Candidates must convince the examiner.

(ii) $4 \times 3^3 = 108$ so $x = 3$

When $x = 3$ then $y = 19$

e This correct response earns both marks.

(c) (i) When $x = 1$, $4(1) + 1 + \dfrac{54}{1^2} = 59$

e The working and answer are fully correct — the mark is awarded.

(ii) Gradient of tangent $= 4 - 108(1)^{-3} = -104$

So gradient of normal $= \dfrac{1}{104}$

Using $y - y_1 = m(x - x_1)$

So $y - 19 = \dfrac{1}{104}(x - 1)$

e The candidate makes one slight slip — with so much in this question he/she has used the incorrect y value (wrong point — 59 should have been used). Therefore, he/she loses 1 mark and earns only 3 marks.

(d) (i) $\int\left(4x + 1 + \dfrac{54}{x^2}\right)dx = \int\left(4x + 1 + 54x^{-2}\right)dx$

$= 2x^2 + x - 54x^{-1} + c$

e All 3 marks are awarded. The working is clear.

(ii) Area below the curve

$$= \left[2x^2 + x - 54x^{-1} \right]_1^3$$

$$= \left[2(3)^2 + 3 - 54(3)^{-1} \right] - \left[2(1)^2 + 1 - 54(1)^{-1} \right]$$

$$= 129 - 57$$

$$= 72$$

Trapezium $= \frac{1}{2}(2)(19 + 59) = 78$

So area $= 78 - 72 = 6$

e This is a good attempt. There is one error in the evaluation of the brackets. The correct answer is 54, making the required area $78 - 54 = 24$. The candidate scores 3 marks and loses the final accuracy mark.

e **The candidate's total mark is 16 out of 19 — just about grade-A category. Many parts show good understanding, but some minor slips have cost marks.**

Quick test

(1) Given that $2\log a^2 - \frac{1}{2} \log b^4 = \log k$, then k is equal to:

  A $a^4 b^2$

  B $\dfrac{a^4}{b^2}$

  C $\dfrac{b^2}{a^4}$

  D $a^2 b^4$

(2) The coefficient of x^2 in the expansion of $(1 + x)(1 + 2x)^5$ is:

  A 15

  B 45

  C 50

  D 10

(3) The expression $3\sin^2 \theta - 4\cos^2 \theta + 4$ can be rewritten as:

  A $7 - \cos^2 \theta$

  B $1 - \cos^2 \theta$

  C $8 - \sin^2 \theta$

  D $7 - 7\cos^2 \theta$

(4) A sector of a circle has an arc length of 12 cm and a radius of 6 cm. The area of the sector is:

  A 9

  B 72

  C 36

  D 12

(5) The sum to infinity of the sequence $32, -8, 2, \ldots$ is:

  A $\dfrac{128}{5}$

  B $\dfrac{1}{2}$

  C Does not exist

  D $\dfrac{128}{3}$

(6) In the sequence defined by $4u_{n+1} = 18 - \frac{1}{2} u_n$, the limiting value L is equal to:

  A 4

  B 18

  C 9

  D $\dfrac{36}{7}$

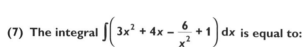

(7) The integral $\int\left(3x^2 + 4x - \dfrac{6}{x^2} + 1\right) dx$ is equal to:

A $6x + 4 + \dfrac{12}{x^3} + c$

B $x^3 + 2x^2 + x + \dfrac{2}{x^3} + c$

C $\dfrac{x^3}{3} + \dfrac{x^2}{2} + \dfrac{6}{x} + c$

D $x^3 + 2x^2 + x + \dfrac{6}{x} + c$

(8) The first term of an arithmetic series is 10, the last term is 100 and the common difference is 5. The sum of this series is equal to:

A 1045

B 990

C 935

D 1140

(9) The curve $y = 1 + 2(3^x)$ is translated by $\begin{bmatrix} 1 \\ -2 \end{bmatrix}$. The new equation is:

A $y = -1 + 2(3^{x+1})$

B $y = 2(3^{x-2})$

C $y = -1 + 2(3^{x-1})$

D $y = 1 + 2(3^{x+1})$

(10) The equation of the normal to the curve $y = 1 + \dfrac{12}{x^3}$ at the point $(1, 13)$ is:

A $y + 36x - 49 = 0$

B $36y - x - 467 = 0$

C $y + 6x - 19 = 0$

D $6y - x - 95 = 0$

Answers

(1) B

(2) C

(3) D

(4) C

(5) A

(6) A

(7) D

(8) A

(9) C

(10) B